ENCYCLOPEDIA OF
ELECTRICAL ENGINEERING

ENCYCLOPEDIA OF
ELECTRICAL ENGINEERING

Edited by
JOHN WILSON

2007

SBS Publishers & Distributors Pvt. Ltd.
New Delhi

ISBN10 : 81-89741-54-3
ISBN13: 978-81-89741-54-9

Price - Rs 795

First Published in India in 2007

© Reserved

Published by:
SBS PUBLISHERS & DISTRIBUTORS PVT. LTD.
2/9, Ground Floor, Ansari Road, Darya Ganj,
New Delhi - 110002, INDIA
Tel: 23289119, 41563911
Email: mail@sbspublishers.com

Printed in India by SALASAR IMAGING SYSTEMS.

Preface

The Encyclopedia on ELECTRICAL ENGINEERING presents a comprehensive list of terms used in the field of ELECTRICAL ENGINEERING and various topics related with it.

This encyclopedia presents a comprehensive list of terms used in the Electrical Engineering and various topics related with it. This is a natural expression of the author's background, training, knowledge and interest. This crucial reference offers a wealth of essential information in a handy, expedient, quick-find layout.

Presented in the arrangement of an encyclopedia, and written in lucid, simple language clear to the general reader, this encyclopedia offers a wealth of information in a portable, convenient, and quick–find format. It includes words, phrases, acronyms and other abbreviations that are used by those who study and write in these fields. The words may be either those used uniquely in the field or more common words that have a special meaning in the context of Electrical Engineering.

The encyclopedia is an excellent reference tool for Students, Educators, Engineers, and equipment manufacturers.

A conscious effort is made to emphasize these basic principles while also providing students with a viewpoint of how computational tools are used in engineering practice.

The style being easy to read, nonnative English Speakers and translators with no engineering know-how will also find the Encyclopedia of ELECTRICAL ENGINEERING very useful.

Abrasive

A hard and wear-resistant material that is used to wear, grind or cut away other material.

Absolute Value

Value of an expression without regard to sign or phase angle.

Absolute Zero

The temperature where thermal energy is at minimum. –273.15°C or 0 Kelvin.

Absorbed Glass Mat (AGM)

A newer type of battery construction that uses saturated absorbant glass mats rather than gelled or liquid electrolyte. Somewhat more expensive than flooded (liquid), but offers very good reliability.

Absorber

In a , the material that readily absorbs photons to generate charge carriers (free electrons or holes).

Absorption Coefficient

The factor by which photons are absorbed as they travel a unit distance through a material.

Acceleration

Rate of change of velocity. [Unit: m/s²]

Accelerometer

A sensor or transducer for measuring acceleration.

Accent Lighting

Directional lighting to emphasize a particular object or draw attention to a part of the field of view.

Acceptor

A dopant material, such as boron, which has fewer outer shell electrons than required in an otherwise balanced crystal structure, providing a hole, which can accept a free electron.

Accumulator

A rechargeable battery or cell.

Accuracy

Specifies the nearness of the measured value from the true value.

Activated Shelf Life

The period of time, at a specified temperature, that a charged battery can be stored before its capacity falls to an unusable level.

Activation Voltage(s)

The voltage(s) at which a charge controller will take action to protect the batteries.

Active Element

An element capable of generating electrical energy.

Active Filter

Any filter using an op amp is called an active filter.

Active Material (Battery)

Material which reacts chemically to produce electrical energy when the cell discharges. The material returns to its original state during the charging process.

Active Power

A term used for power when it is necessary to distinguish among Apparent Power, Complex Power and its components, and Active and Reactive Power.

Active Solar Heater

A solar water or space-heating system that moves heated air or water using pumps or fans.

Actual Capacity or Available Capacity

The total battery capacity, usually expressed in ampere-hours or milliampere-hours, available to perform work. The actual capacity of a particular battery is determined by a number of factors, including the cut-off voltage, discharge rate, temperature, method of charge and the age and life history of the battery.

Actual Peak Load Reductions

Reduction in annual peak load by consumers.

Actuator

Mechanical part of a limit switch that uses mechanical force to actuate the switch contacts.

Adapter

An accessory used for interconnecting non-mating devices or converting an existing device for modified use.

Address

The number that uniquely identifies the location of a word in memory.

Adhesive

A substance that bonds together the surfaces of two other materials.

Adiabatic

Process taking place without heat entering or leaving the system.

Adjustable Set Point

A feature allowing the user to adjust the voltage levels at which a charge controller will become active.

Admittance

Reciprocal of impedance. Ratio of the electric current to the voltage. [Unit: siemens or S]

Advanced Ceramic

A value-added technical ceramic.

Advanced Configuration and Power Interface (ACPI)

An industry-standard specification (co-developed by Hewlett-Packard, Intel, Microsoft, Phoenix, and Toshiba, for operating-system-directed power management for laptop, desktop, and server computers. A replacement for APM.

Advanced Power Management (APM)

Power management standard for computers that provides five power states: Ready, Stand-by, Suspended, Hibernation, Off.

Advanced Product Quality Planning (APQP)

System developed by the AIAG automotive organization to communicate common product quality planning and control plan guidelines for suppliers to the automotive industry.

Aeolian Vibration

A natural forced vibration caused by wind flowing over a conductor. This occurs at alternate wind induced vortices and at wind speeds typically at 8 to 12 MPH. Contact Young and Company for additional information including the formula to calculate Aeolian Vibration.

Aerial Cable
An assembly of insulated conductors installed on a pole or similar overhead structures. It may be self supporting or attached to a messenger cable.

Aerodynamics
Study of air moving around solid objects, or air flowing around a stationary structure.

Affiliated Power Producer
A generating company that is affiliated with a utility.

Aggregation
The process of organizing small groups, businesses or residential customer into a larger, more effective bargaining unit that strengthens their purchasing power with utilities.

Air Blast Breakers
A variety of high voltage circuit breakers that use a blast of compressed air to blow-out the arc when the contacts open. Normally, such breakers only were built for transmission class circuit breakers.

Air Conditioning
The control of temperature, humidity and the purity of the air. In tropical countries like Sri Lanka, air conditioning only cools the air and not heats it, but in cold countries both modes are available.

Air Discharge
A method for testing ESD-protection structures in which the ESD generator is discharged through an air gap between the generator and the device under test (DUT).

Air Leakage Rating
The air leakage rating is a measure of how much air leaks through the crack between the window sash and frame. The rating reflects the leakage from a window exposed to a 25-mile-per-hour wind, and is measured in cubic feet per minute per linear foot of sash crack.

Air Mass
A measure of how far light travels through the Earth's atmosphere. One air mass, or AM1, is the thickness of the Earth's atmosphere. Air mass zero (AM0) describes solar irradiance in space,

where it is unaffected by the atmosphere.

Air Set Cement

A cement that sets through loss of water.

Airfoil

The cross section profile of the leeward side of a wind generator blade. Designed to give low drag and good lift. Also found on an airplane wing.

Alarm

A signal for attracting attention to some abnormal event.

Algebraic Sum

Total of a number of quantities of the same kind, with due regard to sign.

Algorithm

A systematic mathematical procedure which enables a problem to be solved in a definite number of steps.

Aliasing

In A/D conversion, the Nyquist principle states that the sampling rate must be at least twice the maximum bandwidth of the analog signal. If the sampling rate is insufficient, then higher-frequency components are "undersampled" and appear shifted to lower-frequencies. These frequency-shifted components are called aliases.

The frequencies that shift are sometimes called "folded" frequencies because a spectral plot looks like it was folded to superimpose the higher frequency components over the sub-Nyquist portion of the band.

Alkaline

Containing an excess of hydroxyl ions over hydrogen ions.

Alley Arm

A side brace for a crossarm that is not loaded (balanced) evenly.

Alley Roadway (Lighting)

Narrow public ways within a block, generally used for vehicular access to the rear of abutting properties.

Alligator

A speciallized tool attached to a hot stick used to tie a wire or cable into an insulator.

All-or-Nothing Relay

An electrical relay which is

intended to be energized by a quantity, whose value is either higher than that at which it picks up or lower than that at which it drops out.

Alloy
A metal formed by the combination of two or more metals.

Alphanumeric
The collection of numbers, alphabetic characters and symbols.

Alternate Current (AC) Potentiometer
An apparatus for the comparison of a.c. voltages. Balance requires both the magnitude and the phase angle of the unknown voltage to be balanced with the known voltage. This may be done either in cartesian form or in polar form.

Alternate Fuel Vehicle (AFV)
A vehicle powered by fuel other than gasoline or diesel. Examples of alternative fuels are electricity, hydrogen, and CNG.

Alternate Sweep
A vertical mode of operation for a dual-trace oscilloscope. The signal from the second channel is displayed after the signal from the first channel. Each trace has a complete trace, and the display continues to alternate.

Alternating Copolymer
A polymer, composed of two different repeating mers, in which the different mer units systematically alternate positions along the molecular chain.

Alternating Current (AC)
A current whose instantaneous values reverses in regularly recurring intervals of time and which has alternative positive and negative values, the cycle being repeated continuously. The term is commonly used to refer to sinusoidal waveforms.

Alternating Voltage
A voltage which periodically changes its polarity.

Alternation
One-half of a cycle, consisting of the complete rise and fall of an alternating voltage or current in one direction.

Alternator

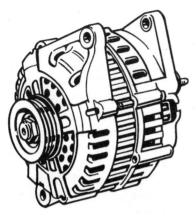

A device that produces alternating current (AC) electricity from the rotation of a shaft. Used in wind and water turbines to generate electricity.

Aluminum Conductor Steel Supported (ACSS)

This is a conductor that is generally used for overhead transmission construction. ACSS is often preferred over ACSR because of its superior sag characteristics.

Ambient Temperature

The temperature of the air, water, or surrounding earth. Conductor ampacity is corrected for changes in ambient temperature including temperatures below 86?F. The cooling effect can in-

crease the current carrying capacity of the conductor. (Review Section 310-10 of the Electrical Code for more understanding)

American Wire Gauge (AWG)

A standard system for designating the size of electrical wire. The higher the number, the smaller the wire. Most house wiring is #12 or 14. In most other countries, wire is specified by the size in millimeters.

Ammeter

Measures the current flow in amperes in a circuit. An ammeter is connected in series in the circuit.

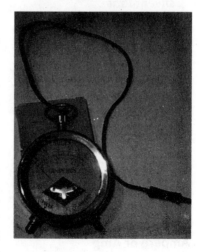

Ampere Hour

A unit of measurement of a battery's electrical storage capacity. Current multiplied by time in hours equals ampere-hours. One amp hour is equal to a current of one ampere flowing for one hour. Also, 1 amp hour is equal to 1,000 mAh

Amperage

This is a name sometimes used in place of current. It is used because the electrical current is measured in Amperes (Amps). By definition, 1 Ampere = the current that will cause silver to be deposited at a rate of 0.001118 grams per second when passed through a solution of silver nitrate.

Amperage Interrupt Capability

Direct current fuses should be rated with a sufficient AIC to interrupt the highest possible current.

Ampere Hour Capacity

The quantity of electricity measured in ampere-hour which may be delivered by a cell or battery under specified conditions.

Ampere or Amp

An Ampere or an Amp is a unit of measurement for an electrical current. One amp is the amount of current produced by an electromotive force of one volt acting through the resistance of one ohm. Named for the French physicist Andre Marie Ampere. The abbreviation for Amp is A but its mathematical symbol is "I". Small currents are measured in milliAmps or thousandths of an Amp.

Ampere Turn (AT)

The product of the number of turns of wire in an electromagnetic coil winding and the current in amperes passing through the winding. This is a direct measure of a reed contact's sensitivity.

Ampere-Hour (Ah)

Quantity of electricity or measure of charge. How many amps flow or can be provided over a one hour period. Most batteries are rated in AH.

Ampere-hour capacity (storage battery)

The number of ampere-hours that can be delivered under specified conditions of temperature, rate of discharge, and final voltage.

Ampere-hour meter

An electricity meter that measures and registers the integral, with respect to time, of the current of a circuit in which it is connected.

Ampere-Turn (AT)

Formerly used as the unit of magnetomotive force (mmf). It is the product of the number of turns in a coil and the current in amperes which flows through it. Since turns is not a unit, the SI Unit of mmf is the ampere]

Amplification

Procedure of expanding the strength of a signal.

Amplifier

A device or circuit used to increase the power current, and voltage level of a signal.

Amplitude

Maximum or peak value of a quantity or wave varying in an oscillatory manner, measured with respect to the reference.

Amplitude Modulation (AM)

A modulation method in which the carrier amplitude changes with the input signal amplitude.

Analog

A system in which an electrical value (usually voltage or current, but sometimes frequency, phase, etc.) represents something in the physical world. The electrical signal can then be processed, transmitted, amplified, and finally, transformed back into a physical quality.

For example: A microphone produces a current that is proportional to sound pressure. Various stages amplify, process, modulate, etc. Ultimately, a varying voltage is presented to a speaker which converts it back to sound waves.

By contrast, a digital system handles a signal as a stream of numbers.

Analog Switch

Switching device capable of switching or routing analog signals (meaning signals that can have any level within a specified legal range), based on the level of a digital control signal.

As a simple example, an audio signal could be switched on or off based on a MUTE signal. Most commonly implemented using CMOS technology integrated circuits. Maxim makes hundreds of examples.

Analog Temperature Sensor

Temperature sensor with a continuous analog voltage or current output that is related, usually linearly, to the measured temperature.

Analog to Digital Converter, A/D converter, A to D Converter

A device or circuit used to convert an analog signal to a digital signal across a pair of terminals.

Analogue Meter

Show a particular (continuous variable) deflection for a given input quantity.

Analogy

A likeness in some ways between dissimilar things that are otherwise unlike. Because of the similarity, many of the equations are identical except for a change of variables or subscripts.

Anchor

A device that supports and holds in place conductors when they are terminated at a pole or structure. The anchor is buried and attached to the pole by way of guy wire to counteract the mechanical forces of these conductors.

Ancillary Services

Necessary services that must be provided in the generation and delivery of electricity. As defined by the Federal Energy Regulatory Commission, they include coordination and scheduling services (load following, energy imbalance service, control of transmission congestion); automatic generation control (load frequency control and the economic dispatch of plants); contractual agreements (loss compensation service); and support of system integrity and security (reactive power, or spinning and operating reserves).

AND Gate

A digital logic circuit used to implement the AND operation. The output of this circuit is 1 only when each one of its inputs is a 1.

Anemometer

A device that measures wind speed. Important in designing a properly sized wind power system.

Angle

A plug that allows the attached flexible cord to exit at right angles.

Angle of Attack

The angle of relative air flow to the blade chord.

Angle of Incidence

The angle that a ray of sun makes with a line perpendicular to the surface. For example, a surface that directly faces the sun has a solar angle of incidence of zero, but if the surface is parallel to the sun (for example, sunrise striking a horizontal rooftop), the angle of incidence is 90°.

Angstrom

A unit used to measure very small lengths, such as wave length. Equal to 10^{-10} m

Angular Velocity (ω)

Rate of rotation about an axis. It is the rate of change of angle with time. It is measured either in revolutions per second, revolutions per minute (r.p.m.) or radians per second (rad/s).

Anisotropic

Exhibiting different values of a property in different crystallographic directions.

Annealing

A generic term used to denote a heat treatment wherein the microstructrure and, consequently, the properties of a material are altered. Frequently, refers to heat treatment whereby a cold-worked metal is softened by allowing it to recrystallize.

Annual Effects

The total effects in energy use (measured in megawatthours) and peak load (measured in kilowatts) caused by all participants in the DSM programs that are in effect during a given year. It includes new and existing participants in existing programs (those implemented in prior years that are in place during the given year) and all participants in new programs (those implemented during the given year). The effects of new participants in existing programs and all participants in new programs should be based on their start-up dates (i.e., if participants enter a program in July, only the effects from July to December should be reported). If start-up dates are unknown and cannot be reasonably estimated, the effects can be annualized (i.e., assume the participants were initiated into the program on January 1 of the given year). The Annual Effects should consider the useful life of efficiency measures, by accounting for building demolition, equipment degradation and attrition.

Annual Equivalent

An equal cash flow amount that occurs every year.

Annual Fuel Utilization Efficiency

A measure of heating efficiency, in consistent units, determined by applying the federal test method for furnaces. This value is intended to represent the ratio of heat transferred to the conditioned space by the fuel energy supplies over one year.

Annual Maximum Demand

The greatest of all demands of the electrical load which occurred during a prescribed interval in a calendar year.

Annual Solar Savings

The annual solar savings of a solar building is the energy savings attributable to a solar feature relative to the energy requirements of a nonsolar building.

Annual Transmission Costs

The total annual cost of the Transmission System shall be the amount specified in Schedule 1 until amended by the Transmission Provider or modified by the Commission.

Annuity

A series of equal cash flows over a number of years.

Anode

During discharge, the negative electrode of the cell is the anode. During charge, that reverses and the positive electrode of the cell is the anode. The anode gives up electrons to the load circuit and dissolves into the electrolyte.

Anodizing
Any electrolytic or chemical process by which a protective or decorative film is released on a metal surface.

Antenna
A device consisting of spaced elements that are used to receive broadcast signals.

Antenna Gain
An antenna's transmission power, provided as a ratio of its output (send) signal strength to its input (recieve) signal strength, normally expressed in dBi. The higher the dBi, the stronger the antenna.

Anthropogenic
Referring to alterations in the environment due to the presence or activities of humans.

Anti-Aliasing
An anti-aliasing filter is used before A/D conversion. It is a lowpass filter that removes signal components

above the Nyquist frequency, thereby eliminating their sampled replicas (aliases) in the baseband.

Antiferromagnetism
A phenomenon observed in some materials in which complete magnetic moment cancellation occurs as a result of antiparallel coupling of adjacent atoms or ions. The macroscopic solid possesses no net magnetic moment.

Anti-Pumping Device
A feature incorporated in a Circuit Breaker or re-closing scheme to prevent repeated operation where the closing impulse lasts longer than the sum of the relay and CB operating times.

Antireflection Coating
A thin coating of a material applied to a solar cell surface that reduces the light reflection and increases light transmission.

Aperture
Opening. In optical instruments, it is the size of the opening admitting light.

Apparent Power (volt-amps)
The product of the applied voltage and current in an ac

circuit. Apparent power, or volt-amps, is not the true power of the circuit because the power factor is not considered in the calculation.

Appliance

Utilization equipment, generally other than industrial, normally built in standardized sizes or types, that is installed or connected as a unit to perform one or more functions such as clothes washing, air conditioning, food mixing, deep frying, etc.

Appliance Saturation

The percentage of households or buildings in a service area that have the type of equipment to which the demand-side technology applies. For example, if 50 percent of the residential customers have a central air conditioner, the appliance saturation is 50 percent.

Applicability Factor

The percentage of end-use energy and demand used by a technology to which the demand-side management (DSM) measure applies. For example, the high-efficiency fluorescent lighting DSM measure applies to fluores-cent lighting but not all lighting. Applicability therefore represents the percent of the lighting end-use attributable to fluorescence for which there could be high-efficiency replacements installed.

Application Program Interface (API)

A software layer that allows a system to be programmed via a defined set of commands.

Aqueous Batteries

Batteries with water-based electrolytes. The electrolyte may not appear to be liquid since it can be absorbed by the battery's separator.

Arc

Sparking that results when undesirable current flows between two points of differing potential. This may be due to leakage through the intermediate insulation or a leakage path due to contamination.

Arc Flash

An arcing fault is the flow of current through the air between phase conductors or phase and neutral or ground. An arcing fault can release tremendous amounts of concentrated radiant energy at the point of the arcing in a small fraction of a second resulting in extremely high temperatures, a tremendous pressure blast, and shrapnel hurling at high velocity.

Arc Thermal Performance Value

Maximum capability for arc flash protection of a particular garment or fabric measured in calories per square centimeter. Though both garments and fabrics can be used for protection a garment made from more than one layer of arc flash rated fabric will have a calorie per square centimeter rating greater than the sum of the ATPV rating of the original fabrics. The calorie per square centimeter rating of most arc flash protective suits, coveralls, and coats is commonly sewn into the fabric in large letters.

Arc-fault Circuit Interrupter (AFCI)

A breaker that shuts off current in a circuit instantly when an arc fault is detected. Arcing Time
The time between instant of

separation of the CB contacts and the instant of arc excitation.

Arc-over Voltage

The minimum voltage required to cause an arc between electrodes separated by a gas or liquid insulation.

Are

A metric unit of area, especially with land. Equal to 100 square meter.

Area Load

The total amount of electricity being used at a given point in time by all consumers in a utility's service territory.

Argand Diagram

A plot of the cartesian co-ordinates (real and imaginary components) or the polar co-ordinates of a complex number on the x-y plane.

Arithmetic Progression

A series of quantities in which each term differs from the preceding term by a constant difference.

Armature

The coil or coils of an electric motor or generator or of an electric apparatus in which a voltage is induced by a magnetic field.

Armature Coil

A winding that develops current output from a generator when its turns cut a magnetic flux.

Armor

An outer metal layer applied to a cable for mechanical protection. Armor is comprised of factory formed wire, designed to be applied to a range of conductor sizes. Preformed Line Products manufacturers Armor.

Armor Rod

An outer metal layer applied to a cable for mechanical protection. Armor Rods are comprised of factory formed wire, designed to be applied to a range of conductor sizes. Preformed Line Products manufacturers Armor Rods.

Armored Cable

This cable contains two insulated conductors and a thin aluminum or copper bonding strip inside a metal sheath. The metal sheath is the ground, not the bonding strip.

Arm's Reach

A zone of accessibility to touch, extending from any point on a surface where persons usually stand or move about to the limits which a person can reach with a hand in any direction without assistance.

Array

Any number of photovoltaic modules connected together to provide a single electrical output. Arrays are often designed to produce significant amounts of electricity.

Array Operating Voltage

The voltage produced by a photovoltaic array when exposed to sunlight and connected to a load.

Arrester

A nonlinear device to limit the amplitude of voltage on a power line. The term implies that the device stops overvoltage problems (i.e. lightning). In actuality, voltage clamp levels, response times and installation determine how much voltage can be removed by the operation of an arrester.

Ash

Impurities consisting of

silica, iron, alumina, and other noncombustible matter that are contained in coal. Ash increases the weight of coal, adds to the cost of handling, and can affect its burning characteristics. Ash content is measured as a percent by weight of coal on an "as received" or a "dry" (moisture-free, usually part of a laboratory analysis) basis.

Askeral

A generic term for a group of synthetic, fire-resistant, chlorinated aromatic hydorcarbons used as electrical insulated fluids.

Asset

An economic resource, tangible or intangible, which is expected to provide benefits to a business.

Asymmetric

Not possessing symmetry. Unequal distribution about one or more axes.

Atactic

A type of polymer chain configuration wherein side groups are randomly poitioned on one side of the polymer backbone or the other.

Atmosphere

Unit of pressure corresponding to standard atmospheric pressure. It is taken as the pressure that will support a column of mercury 760 mm high. It is also equal to $1.013 \ 10^5$ pa

Atomic Mass Unit (amu)

Unit used for expressing masses of isotopes of elements. 1 a.m.u. = 1.661×10^{-27} kg

Attenuation

The reduction of a signal from one point to another. For an electrical surge, attenuation refers to the reduction of an incoming surge by a limiter (attenuator). Wire resistance, arresters, power conditioners attenuate surges to varying degrees.

Attenuator

A passive device used to reduce signal strength.

Atto (a)

Decimal sub-multiple prefix corresponding to 10^{-18}

Attributes

Attributes are the outcomes by which the relative "goodness" of a particular expansion plan is measured e.g. fuel usage. Some attributes, such as fuel usage, are measured in well-defined parameters. Other attributes (e.g. public perception of a technology) are more subjective. Attributes may be grouped in several ways. Categories include financial, economic, performance, fuel usage, environmental, and socio-economic. The attributes chosen must measure issues that directly concern the utility and have an impact on its planning objectives. Limiting the number of attributes reduces the complexity and cost of a study.

Austenite

Face-centered cubic iron; also iron and steel alloys that have the FCC structure.

Automated Protection System

A system of protective devices that automatically opens a circuit when the current exceeds a set limit.

Automatic Gain Control (AGC)

A circuit that modulates an amplifier's gain, in response to the relative strength of the input signal, in order to maintain the output power.

Automatic line sectionalizer

A self-contained circuit-opening device that automatically opens the main electrical circuit after sensing and responding to a predetermined number of successive main current impulses.

Automatic Meter Reading (AMR)

A system installed to read a utility meter remotely.

Automatic Power Control (APC)

Feature in laser drivers that uses feedback from the laser to adjust the drive, to keep the laser's output constant.

Automatic Recloser

An automatic switch used to open then reclose following an over current event on a distribution voltage (medium voltage) line.

Automatic Reset

A starter that automatically restarts a new replacement fluorescent lamp after the circuit is energized.

Automatic Transfer Switch

A switch that automatically transfers electrical loads to alternate or emergency-standby power sources.

Automation

The application of mechanical or electronic techniques to minimise the use of manpower in any process.

Autonomous System

A stand-alone PV system that has no back-up generating source. May or may not include storage batteries. Most battery systems are designed for a certain minimum "days of autonomy" — which means that the batteries can supply sufficient power with no sunlight to charge the batteries. This varies from 3-5 days in the sunbelt, to 5 to 10 days elsewhere.

Autoreclose

A feature of certain circuit breakers where they close automatically after a predetermined time after an automatic opening due to a transient fault.

Autotransformer

An autotransformer is a transformer that uses a common winding for both the primary and secondary windings. Essentially an inductor with a center-tap, an autotransformer is often used in power-supply boost-converter applications to achieve a higher output voltage, while limiting the peak flyback voltage seen by the power switch.

Autotransformer Starter

A starter that includes an auto-transformer to furnish reduced voltage for starting an alternating current motor.

Auxiliary Contacts

The contacts of a switching device, in addition to the main current contacts, that operate with the movement of the latter. They can be normally open (NO) or normally closed (NC) and change state when operated.

Auxiliary Power

The power required for correct operation of an electrical or electronic device, supplied via an external auxiliary power source rather than the line being measured.

Auxiliary Relay

An all-or-nothing relay energized via another relay. An example is a measuring relay, for the purpose of providing higher rated contacts, or introducing a time delay, or providing multiple outputs from a single input.

Auxiliary Source

A power source dedicated to providing emergency power to a critical load when commercial power is interrupted.

Avalanche

A build up of particles caused by the collision of a high energy particle with any other form of matter. [Note: The term is derived from the avalanches occurring in a mountain]

Avalanche Photo Diode (APD)

A photodiode designed to take advantage of avalanche multiplication of photocurrent to provide gain. As the reverse-bias voltage approaches the break-down voltage, hole-electron pairs created by absorbed pho-

tons acquire sufficient energy to create additional hole-electron pairs when they collide with ions. Thus a multiplication or signal gain is achieved.

Average Revenue per Kilowatthour

The average revenue per kilowatthour of electricity sold by sector (residential, commercial, industrial, or other) and geographic area (State, Census division, and national), is calculated by dividing the total monthly revenue by the corresponding total monthly sales for each sector and geographic area.

Avoided Energy Cost

Avoided energy cost of fuel and operating and maintaining utility power plants.

Azimuth

Angle between the north direction and the projection of the surface normal into the horizontal plane; measured clockwise from north. As applied to the PV array, 180 degree azimuth means the array faces due south.

Back Emf

The emf set up in the coil of an electric motor, opposing the current flowing through the coil, when the armature rotates.

Back Flashover

Flash-over occurring from an object usually at earth potential (such as a tower) to a line conductor due to the potential of the earthed object rising due to lightning.

Backup Step-Up

Step-up, switching-regulator power supply with a backup battery switchover.

Bainite

A Fe-C composition consisting of a fine dispersion of cementite in alpha-ferrite. It is an austenitic transformation product that forms at temperatures between those at which pearlite and martensite transformations occur.

Baker Board

A platform used to work above the ground on a wood pole.

Balance of System (BOS)

Represents all components and costs other than the PV modules. It includes design costs, land, site preparation, system installation, support structures, power conditioning, operation and maintenance costs, batteries, indirect storage, and related costs.

Balanced Load

Refers to an equal loading of the phases in a polyhphase system (current and phase angle).

Balanced Polyphase System

A polyphase system in which both the currents and voltages are symmetrical.

Balanced Three Phase

A three phase voltage or current is said to be balanced when the magnitude of each phase is the same, and the phase angles of the three phases differ from each other by 120 A star-connected load or a delta-connected load is said to be balanced when the three arms of the star or the delta have equal impedances in magnitude and phase.

Balancing

With wind turbine blades, adjusting their weight and weight distribution through 2 axes so that all blades are the same. Unbalanced blades create damaging vibration.

Ballast

A device that by means of inductance, capacitance, or resistance, singly or in combination, limits the lamp current of a fluorescent or high intensity discharge lamp. It provides the necessary circuit conditions (voltage, current and wave form) for starting and operating the lamp.

Ballistic Galvanometer

Instrument for measuring the total quantity of electricity passing through a circuit due to a momentary current. The period of oscillation of the galvanometer must be long compared with the time during which the current flows.

Ballistics

The study of the flight path of projectiles.

Band Gap

In a semiconductor, the energy difference between the highest valence band and the lowest conduction band.

Band Gap Energy (Eg)

The amount of energy (in electron volts) required to free an outer shell electron from its orbit about the nucleus to a free state and, thus, to promote it from the valence level to the conduction level.

Bandpass Filter

A filter designed to pass all frequencies within a band of frequencies.

Bandstop Filter

A filter designed to eliminate all frequencies within a band of frequencies.

Band-To-Band Auger Recombination

Recombination of an electron and a hole occurring between bands of the same energy in which no magnetic radiation is emitted.

Bank

A group of electrical devices, usually transformers or capacitors, connected in a way to increase capacity.

Bar

Unit of pressure equivalent to 10^5 pa.

Bare Conductor

A conductor not covered with insulating material.

Barier

A part providing a defined degree of protection against contact with live parts from any usual direction of access.

Barometer

Instrument for measuring atmospheric pressure.

Barier Energy

The energy given up by an electron in penetrating the cell barrier; a measure of the electrostatic potential of the barrier.

Base

The part of a transistor which separates the emitter from the collector. The middle part of the transistor. permits electrons from emitter to pass through to the collector.

Base Bill

A charge calculated through multiplication of the rate from the appropriate electric rate schedule by the level of consumption.

Base Line

A vibration reading taken when a machine is in good operating condition that is used as a reference for monitoring and analysis.

Base Load

The minimum load experienced by an electric utility system over a given period of time, which must be supplied at all times.

Base Load Capacity

Capacity of generating equipment operated to serve loads 24-hours per day.

Base Load Plant

A power plant built to operate around-the-clock. Such plants tend to have low operating costs and high capital costs and are best utilized by running continuously. Coal fired and nuclear fuelled plants are typical base load plants.

Base Load Unit

A generating unit that normally operates at a constant output to take all or part of the base load of a system.

Base Rate

The portion of the total electric or gas rate covering the general costs of doing business unrelated to fuel expenses.

Base Transceiver Station (BTS)

The stationary component of a cellphone system includes transmit-receive units and one or more antennae. The combined systems (often including multiple co-located systems and ganged directional antennae) is called a cell-site, a base station, or a base transceiver station (BTS).

Base Year

The first year of the period of analysis. The base year does not have to be the current year.

Baseline

The electrical signal from a sensor when no measured variable is present. Often referred to the output at no-load condition.

Baseline Forecast

A prediction of future energy needs which does not take into account the likely effects of new conservation programs that have not yet been started.

Baseline Performance Value

Initial values of Isc, Voc, Pmp, Imp measured by the accredited laboratory and corrected to Standard Test Conditions, used to validate the manufacturer's performance measurements provided with the qualification modules per IEEE 1262.

Baseload

The minimum amount of electric power delivered or required over a given period of time at a steady rate.

Baseload Capacity

The generating equipment normally operated to serve loads on an around-the-clock basis.

Baseload Plant

A plant, usually housing high efficiency steam-electric units, which is normally operated to take all or part of the minimum load of a system, and which consequently produces electricity at an essentially constant rate and runs continuously. These units are operated to maximize system mechanical and thermal efficiency and minimize system operating costs.

Basic Impulse Level (BIL)

A reference impulse (voltage) insulation strength expressed in terms of the peak value of the withstand voltage of a standard impulse voltage wave. It is used to express the ability of electrical equipment such as transformers to withstand certain levels of voltage impulses like lightning strokes.

Basic Insulation

Insulation applied to live parts to provide basic protection against electric shock and other hazards, which do not necessarily include insulation used exclusively for functional purposes.

Basic Insulation Level (BIL)

It defines the insulation level of power system equipment. It is a statement of the impulse (lightning or switching as appropriate) withstand voltage and the short duration power frequency withstand voltage.

Basic Service

The four charges for generation, transmission, distribution and transition that all customers must pay in order to retail their electric service.

Bass Boost

Circuitry that boosts the bass response of the amplifier, improving audio reproduction, especially when using inexpensive headphones.

Battery

Two or more electrochemical cells enclosed in a container and electrically interconnected in an appropriate series/parallel arrangement to provide the required operating voltage and current levels. Under common usage, the term battery also applies to a single cell if it constitutes the entire electrochemical storage system.

Battery Available Capacity

The total maximum charge, expressed in ampere-hours, that can be withdrawn from a cell or battery under a specific set of operating conditions including discharge rate, temperature, initial state of charge, age, and cut-off voltage.

Battery Backup

A battery or a set of batteries in a UPS system. Its purpose is to provide an alter-

nate source of power if the main source is interrupted.

Battery Capacity

The electric output of a cell or battery on a service test delivered before the cell reaches a specified final electrical condition and may be expressed in ampere-hours, watt-hours, or similar units. The capacity in watt-hours is equal to the capacity in ampere-hours multiplied by the battery voltage.

Battery Cell

The simplest operating unit in a storage battery. It consists of one or more positive electrodes or plates, an electrolyte that permits ionic conduction, one or more negative electrodes or plates, separators between plates of opposite polarity, and a container for all the above.

Battery Charger

A device or a system which provides the electrical power needed to keep the battery backup fully charged.

Battery Cycle Life

The number of cycles, to a specified depth of discharge, that a cell or battery can un-

. . . .

dergo before failing to meet its specified capacity or efficiency performance criteria.

Battery Electric Vehicle (BEV)

An EV powered by electricity stored in batteries.
C1
C2
Cn
Cn are a series of n-hour battery charge/discharge test specifications from the SAE. Battery charge/discharge rates affect storage capacity.

Battery Energy Capacity

The total energy available, expressed in watt-hours (kilowatt-hours), which can be withdrawn from a fully charged cell or battery. The energy capacity of a given cell varies with temperature, rate, age, and cut-off voltage. This term is more common to system designers than it is to the battery industry where capacity usually refers to ampere-hours.

Battery Energy Storage

The three main applications for battery energy storage systems include spinning reserve at generating stations, load leveling at substations, and peak shaving on the customer side of the

meter. Battery storage has also been suggested for holding down air emissions at the power plant by shifting the time of day of the emission or shifting the location of emissions.

Battery Freshness Seal

A feature in microprocessor supervisory circuits which disconnects a backup battery from any down-stream circuitry until V_{CC} is applied the first time. This keeps a backup battery from discharging until the first time a board is plugged in and used, and thus preserves the battery life.

Battery Fuel Gauge

A feature or device that measures the accumulated energy added to and removed from a battery, allowing accurate estimates of battery charge level.

Battery Life

The period during which a cell or battery is capable of operating above a specified capacity or efficiency performance level. Life may be measured in cycles and/or years, depending on the type of service for which the cell or battery is intended.

Battery Monitor

A feature that monitors the voltage on a battery and indicates when the battery is low. It is usually implemented using a comparator to compare the battery voltage to a specified level. May also include functions such as charging, remaining capacity estimation, safety monitoring, unique ID, temperature measurement, and nonvolatile (NV) parametric storage.

Battery Switchover

A circuit that switches between the higher of a main supply and a backup battery.

Battery Tray

A contained with a base and walls for holding several cells or batteries.

Battery-Charge Rate

The current expressed in amperes (A) or milli amps (mA) at which a battery is charged.

Bayonet

Designed for incandescent lamps having an unthreaded metal shell with two diametrically opposite keyways that mate with the keyways on the lampholder. Pushing down on the bulb and turning it clockwise in the lampholder locks the bulb in place.

Bay-O-Net

A fusing device frequently used to protect transformers and downstream devices. A Bay-O-Net fuse fuse may include a Partial Range Current Limiting Fuse in series with n under oil fuse link.

Bayonet Coupling

A quick coupling device for plug and receptacle connectors, accomplished by rotation of a cam operating device designed to bring the connector halves together.

Beacon

In wireless networking, a beacon is a packet sent by a connected device to inform other devices of its presence and readiness.

Beam Spread

With regard to outdoor lightg, the angle between the two directions in a plane in which the intensity is equal to a stated percentage of the maximum beam intensity. The percentage is typically 10% for floodlights and 50% for roadway luminaires.

Beckmann Thermometer
Sensitive thermometer for measuring small differences or changes in temperature.

Bel
Measurment of a signal's power compared to a reference; also, measurement of sound pressure.

Belt
Refers to a lineman's climbing belt.

Beryllium Copper (BeCu)
A relatively expensive contact material with properties superior to brass and phosphor bronze. It is recommended for contact applications requiring repeated extraction and reinsertion because of its resistance to fatigue at high operating temperatures.

Bewley Lattice Diagram
This is a convenient diagram devised by Bewley, which shows at a glance the position and direction of motion of every incident, reflected, and transmitted surge on the system at every instant of time. The diagram overcomes the difficulty of otherwise keeping track of the multiplicity of successive reflections at the various junctions.

Bias Current
The current used as a bias quantity in a biased relay.

Biased Relay
A relay in which the characteristics are modified by the introduction of some quantity, and which is usually in opposition to the actuating quantity.

Bidirectional
The device accommodates signals traveling either direction though a single channel.

Bifunctional Monomer
A monomer unit that has two active bonding positions.

Bikeway (Lighting)
Any road, street, path or way that is specifically designated as being open to bicycle travel, regardless of whether such facilities are designed for the exclusive use of bicycles.

Bilateral Agreement
Written statement signed by

a pair of communicating parties that specifies what data may be exchanged between them.

Bilateral Contract

A direct contract between the power producer and user or broker outside of a centralized power pool or power exchange.

Bimetallic Ştrip

A strip composed of two different metals welded together in such a way that a rise of temperature will cause it to deform as a result of the unequal expansion. It is used in switches for control of temperature.

Binary Cell

An element in a computer which can store information by virtue of its ability to remain stable in one of two possible states.

Binary Notation

A system of numbers which has only two different integer values 0 and 1.

Binary-Coded Decimal (BCD)

Representation of a number in which each decimal digit (0-9) is encoded in binary, with four bits per decimal digit.

Biomass

Any plant-derived organic matter available on a renewable basis, including dedicated energy crops and trees, agricultural food and feed crops, agricultural crop wastes and residues, wood wastes and residues, aquatic plants, animal wastes, municipal wastes, and other waste materials.

Bipolar Inputs

An input which accommodates signals both above and below ground.

Bipolar Junction Transistor

A Bipolar Junction Transistor, or BJT, is a solid-state device in which the current flow between two terminals (the collector and the emitter) is controlled by the amount of current that flows through a third terminal (the base).

Contrast to the other main transistor type, the FET, in which the ouput current is controlled by input voltage (rather than by input current).

Bipolar Transistor

A transistor where both free electrons and holes are necessary for normal operation.

Building-Integrated PV

A term for the design and integration of PV into the building envelope, typically replacing conventional building materials. This integration may be in vertical facades, replacing view glass, spandrel glass, or other facade material; into semitransparent skylight systems; into roofing systems, replacing traditional roofing materials; into shading "eyebrows" over windows; or other building envelope systems.

Bisection

Division into two equal parts.

Bit

The unit of information in information theory. The amount of information required to specify one of two alternatives 0 and 1.

Bit Banging

A technique which uses the general-purpose ports of a microcontroller to emulate a serial interface standard (I^2C, SPI, etc).

Bit Error Rate (BER)

A measure of the number of erroneous bits which can be expected in a specified number of bits in a serial stream.

Bit Error Rate (BER) Tester (BERT)

A piece of test equipment which determines the bit error rate for a device under test (DUT).

Bit Error Ratio

The number of erroneous bits divided by the total number of bits transmitted, received, or processed over some stipulated period.

Black Body

Thermal radiator that absorbs completely all incident radiation, whatever the wavelength, the direction of incidence or the polarization.

Blackout

A total loss of the commercial electrical power lasting for more than one cycle. Blackouts can result from any of a number of problems, ranging from Acts of God (high winds, storms, lightning, falling trees, floods, etc.) to situations such as cables being cut during excavation, equipment failures at the utility, vandalism, corrosion, etc. Used synonymously with Outage.

Blade
The part of a wind generator rotor that catches the wind.

Blade Server
A blade server is a computer system on a motherboard, which includes processor(s), memory, a network connection, and sometimes storage. The blade idea is intended to address the needs of large-scale computing centers to reduce space requirements for application servers and lower costs.

Blanking
The process of making the trace, or parts of a trace, invisible.

Bleaching
Removing the colour from coloured materials by chemical transformation.

Blink Control
Controls the display segment blink rate.

Block Copolymer
A linear copolymer in which identical mer units are clustered in blocks along the molecular chain.

Blocking Diode
A semiconductor connected in series with a solar cell or cells and a storage battery to keep the battery from discharging through the cell when there is no output, or low output, from the solar cell. It can be thought of as a one-way valve that allows electrons to flow forwards, but not backwards.

Blower Doors
Energy contractors use blower doors to see how much air leaks through windows, doors, and other places in your house. The blower door is a large board that blocks the front door of your house. A powerful fan installed in the door draws the air out of your house and causes a strong draft inside where ever the air is leaking in. This can help the contractor locate the air leaks, and gives a good overall indication of how "leaky" your house is.

Blowing
The act of installing fiberoptic cable into a duct using air pressure.

Bluetooth
A technology that allows voice and data connections between a wide range of

mobile and stationary devices through short-range digital two-way radio. For instance, it specifies how mobile phones, Wireless Information Devices (WIDs), computers and PDAs interconnect with each other, with computers, and with office or home phones.

Board of Trade Unit or BOT Unit

Unit of electrical energy (British) supplied to the consumer. Equal to 1 kWh. Energy obtained when a power of 1 kW of power is maintained for 1 hour.

Bode Plot

Semi-log plots of the magnitude (in decibel) and phase angle of a transfer function (or performance) against frequency.

Body-centered Cubic (BCC)

A common crystal structure that contains atoms located at the corners of a cubic cell and one atom at the cell center position.

Boiler

A device for generating steam for power, processing, or heating purposes or for producing hot water for heating purposes or hot water supply. Heat from an external combustion source is transmitted to a fluid contained within the tubes in the boiler shell. This fluid is delivered to an end-use at a desired pressure, temperature, and quality.

Boiling

The state of a liquid at its boiling point when the maximum vapor pressure of the liquid is equal to the external pressure to which the liquid is subject, and the liquid is freely converted into vapor.

Boiling Point

The temperature at which the maximum vapor pressure of the liquid is equal to the external pressure. The temperature at which the liquid boils freely under that pressure.

Boiling Water Reactor (BWR)

A nuclear reactor in which water is used as coolant and moderator. Steam is thus produced in the reactor under pressure and can be used to drive a turbine.

Boltzmann's Constant

The gas constant per molecule: 1.381x10-23J/atom K; 1.381x10-16 erg/atom K; or 8.63x10-5 ev/atom K.

Bonding

A complete and permanent electrical connection. The permanent joining of metallic parts to form an electrically conductive path that ensures electrical continuity and the capacity to conduct safely any current likely to be imposed.

Bonding Conductor

A protective conductor providing equipotential bonding.

Bonding Energy

The energy required to separate two atoms that are chemically bonded to each other.

Bonding Jumper

A bare or insulated conductor used to ensure the required electrical conductivity between metal parts required to be electrically connected. Frequently used from a bonding bushing to the service equipment enclosure to provide a path around concentric knockouts in an enclosure wall: also used to bond one race-

way to another.

Boolean Algebra

A branch of symbolic logic used in computers. Logical operations are performed by operators such as "and", "or", in a way analogous to mathematical signs.

Boomer

A lineman that moves from job to job.

Boost Charge

A charge applied to a battery which is already near a state of full charge, usually of short duration.

Boost Converter

A power supply that steps an input voltage up (boosts it) to a higher, regulated voltage.

Booster Transformer

A current transformer whose primary winding is in series with the catenary and secondary winding in the return conductor of a classically-fed A.C. overhead electrified railway.

Bootstrap

Often refers to using the output of a step-up converter to drive the main power FET switch, providing more gate drive than the input can supply alone. Also refers to using a switched capacitor to boost the voltage of a node.

Bottom Ash

Slag or other residue remaining in the boiler after coal is burned.

Boule

A sausage-shaped synthetic single-crystal mass grown in a special furnace, pulled and turned at a rate necessary to maintain the single-crystal structure during growth.

Bounce

Intermittent opening and closing of closed contacts or closing and opening of open contacts, usually implying the motion resulting from contact impact.

Bounce Time (in milliseconds)

Time taken for a bounce.

Box

Device for mounting electrical fixtures and their wiring in walls and ceilings. Common varieties include new work, old cork, single gang, two-gang, fan, and junction.

Bragg's Law

A relationship that stipulates the condition for diffraction by a set of crystallographic planes.

Brake Horse Power (bhp)

Horse power of an engine measured by the degree of resistance offered by a brake. Represents the useful power that the machine can develop.

Braking System

A device to slow a wind turbine's shaft speed down to safe levels electrically or mechanically.

Branch

An element in a circuit connecting two nodes.

Branch Circuit, Appliance

A branch circuit that supplies energy to one or more outlets to which appliances are to be connected, and that has no permanently connected lightning fixtures that are not a part of an appliance.

Branched Polymer

A polymer having a molecular structure of secondary chains that extend from the primary chains.

Brass

A copper-rich copper-zinc alloy.

Break-Before-Make

A switch that is configured to break (open) the first set of contacts before engaging (closing) the new contacts. This prevents the momentary connection of the old and new signal paths.

Breakdown

The occurrence of a large current between electrodes separated by a dielectric at a critical voltage.

Breakdown Torque

The maximum torque that an AC motor will develop with rated voltage applied at rated frequency without an abrupt drop in speed. Also termed pull-out torque or maximum torque.

Breakdown Voltage

The voltage which may be applied between insulated parts of a reed contact without damage, arcing, breakdown , or causing excessive leakage.

Breaker

Short for circuit breaker.

Breeder Reactor

A nuclear reactor which produces the same kind of fissile material as it burns. For example, a reactor using plutonium as a fuel can produce more plutonium than it uses by conversion of Uranium-238.

Bridge Battery

A battery intended to provide power to system memory while the main battery is replaced.

Bridge Rectifier

A full-wave rectifier where the diodes are connected in a bridge circuit (two of them are always conducting at any given time). This allows the current to the load during both the positive and negative alternation of the supply voltage. This is the most common type of rectifier circuit to produce a unidirectional voltage for an alternating input.

Bridge-Tied Load

Used in audio applications, the load (a speaker in this case) is connected between two audio amplifier outputs (it "bridges" the two output terminals).
This can double the voltage

swing at the speaker, compared to a speaker that is connected to ground. The ground-tied speaker can have a swing from zero to the amplifier's supply voltage. A BTL-driven speaker can see twice this swing because the amplifier can drive either the + terminal of the speaker or the—terminal, effectively doubling the voltage swing.

Since twice the voltage means four times the power, this is a major improvement, especially in applications where battery size dictates a lower supply voltage—e.g. automotive or handheld applications.

Brightness

Brightness is the quotient of the luminous intensity of a small element of the source and the area of the element projected on to a plane perpendicular to the given direction. [Unit: candela per unit area or Cd/m^2]

British Thermal Unit (BTU)

The standard unit for measuring heat energy. One Btu is the amount of heat required to raise the temperature of water one degree Farenheit. One kwh = 3,414 Btu's (A burning match will produce approximately 1 Btu of heat.)

Brittle Fracture

Fracture that occur by rapid crack propagation and without appreciable macroscopic deformation.

Brownout

A controlled power reduction in which the utility decreases the voltage on the power lines, so customers receive weaker electric current. Brownouts can be used if total power demand exceeds the maximum available supply. The typical household does not notice the difference.

Brushes

Devices for transferring power to or from a rotating object. Usually made of carbon-graphite. Found in electric motors, alternators and generators.

Buck

A "buck" or "step-down" switch-mode voltage regulator is one in which the output voltage is lower than its input voltage.

Note: A customer asked the origin of the term and no

one seems to know! A buck regulator is a step-down regulator, as opposed to boost. We think it's an American term—in England it was always "step-down." Buck means to resist or reduce (as in "buck the trend"), and hence was used to denote a step-down. Conveniently, it alliterates with the opposite, a boost regulator.

Buck-Boost
A switch-mode voltage regulator in which output voltage can be above or below the input voltage.

Buck-Boost Transformer
A small, low voltage transformer placed in series with the power line to increase or reduce steady state voltage.

Bucket Truck
An aerial lift truck used to lift men high enough to work on overhead lines.

Buckets
In an impulse hydro turbine, the buckets are attached to the turbine near the runner, and used to 'catch' the water. The force of the water hitting the buckets turns the runner,

which turns the alternator drive shaft, causing the alternator to generate power.

Buffer Register
The register that holds digital data temporarily.

Building Wire
Conductors and cables used in commercial building construction.

Building-Integrated Photovoltaics (BIPV)
A term for the design and integration of photovoltaic (PV) technology into the building envelope, typically replacing conventional building materials. This integration may be in vertical facades, replacing view glass, spandrel glass, or other facade material; into semitransparent skylight systems; into roofing systems, replacing traditional roofing materials; into shading "eyebrows" over windows; or other building envelope systems.

Bulb
The outer enclosure of a light source; usually glass or quartz.

Bulk Power Market
Wholesale purchases and sales of electricity.

Bull Line

Heavy line used to pull wire or cable into a conduit or into an overhead configuration.

Bull Wheel

A reel device used to hold tension during the wire installation process.

Bunched Stranding

A term applied to a number of wires twisted together in one direction in one operation without regard to their geometric arrangement.

Bundle

Multiple cables used to form one phase of an overhead circuit.

Bundled Service

Customers receive electric generation, transmission, distribution, and related customer service and support functions as a combined service.

Bundled Utility Service

All generation, transmission, and distribution services provided by one entity for a single charge. This would include ancillary services and retail services.

Burden

Load imposed by an electronic or electrical device on the measured input circuit, expressed in volt-amps.

Burgers Vector

A vector that denotes the magnitude and direction of lattice distortion associated with a dislocation.

Burst Dimming

Burst Dimming is a method of controlling the brightness of cold cathode fluorescent lamps (CCFL) by turning the lamps on and off at a rate faster than the human eye can detect. The on/off rate is nominally 100Hz to 300Hz. The higher the ratio of on-time to off-time, the brighter the lamps will be. Because of CCFL response times, on-time to off-time ratios of less than 1% are not practical.

Burst Mode

1. A temporary high-speed data-transfer mode that can transfer data at significantly higher rates than would normally be achieved with nonburst technology.
2. The maximum short-term throughput which a device is capable of transferring data.

Bus

Data path that connects to a number of devices. A typical example is the bus a computer's circuit board or backplane. Memory, processor, and I/O devices may all share the bus to send data from one to another. A bus acts as a shared highway and is in lieu of the many devoted connections it would take to hook every device to every other device. Often misspelled "buss."

Busbar

A rigid conductor used for connecting together distributors or feeders.

Bushing

Bushings are insulators which are used to take high voltage conductors through earthed barriers such as walls, floors, metal, and tanks.

Butterworth Filter

A filter designed to produce a flat response up to the cutoff frequency.

Buy Through

An agreement between utility and customer to import power when the customer's service would otherwise be interrupted.

Bypass

A capacitor placed from a dc signal to ground will remove any ac component of the signal by creating an ac short circuit to ground.

Bypass Capacitor

A capacitor placed from a dc signal to ground to remove any ac component of the signal by creating an ac short circuit to ground.

Bypass Diode

A diode connected across one or more solar cells in a photovoltaic module such that the diode will conduct if the cell(s) become reverse biased. [UL 1703] Alternatively, diode connected antiparallel across a part of the solar cells of a PV module. It protects these solar cells from thermal destruction in case of total or partial shading, broken cells, or cell string failures of individual solar cells while other cells are exposed to full light.

Cable

A term generally applied to the larger sizes of bare or weatherproofed (covered) and insulated conductors. It is also applied to describe a number of insulated conductors twisted or grouped together.

Cable Coupler

A means of enabling the connection or disconnection, at will, of two flexible cables. It consists of a connector and a plug.

Cable Ducting

An enclosure of metal or insulating material, other than conduit or cable trunking, intended for the protection of cables which are drawn in after erection of the ducting.

Cable Ladder

A cable support consisting of a series of transverse supporting elements rigidly fixed to main longitudinal supporting members.

Cable Pulling Lubricant

A chemical compound used to reduce pulling tension by lubricating a cable when pulled into a duct or conduit.

Cable Sheath

The outermost covering of a cable providing overall protection

Cable Tray

A rigid structural system used to support cables and raceways. Types of cable trays include ladder, ventilated trough, ventilated channel, and solid bottom.

Cable Tunnel

A corridor containing supporting structures for cables and joints and/or other elements of wiring systems and whose dimensions allow persons to pass freely throughout the entire length.

Candela (Cd)

The candela is the SI unit of luminous intensity. It is defined as the luminous intensity, in a given direction, of a source that emits monochromatic radiation of frequency 540 x 1012 hertz and that has a radiant intensity in that direction of 1/683 watt per steradian [1979]

Candelabra

A small screw-base threaded lampholder designed for candelabra-base incandescent lamps commonly used in chandeliers, night lights, and ornamental lighting.

Candle Power

The candle power of a light source, in a given direction, is the luminous intensity of the source in that direction expressed in terms of the candela.

Candlepower Distribution Curve

A curve, generally polar, representing the variation of luminous intensity of a lamp or luminaire in a plane

through the light center.

Capability

The maximum load that a generating unit, generating station, or other electrical apparatus can carry under specified conditions for a given period of time without exceeding approved limits of temperature and stress.

Capacitance

The property of a system of conductors to and dielectrics which enables the system to store electricity when a voltage is applied between the conductors, expressed as a ratio of the electrical charge stored and the voltage across the conductors. The basic unit is the farad .

Capacitive Crosstalk

A phenomenon where a signal on one line/trace is capacitively coupled to an adjacent line/trace.

Capacitive Reactance

The effect of capacitance in opposing the flow of alternating or pulsating current.

Capacitor

A device used to boost the voltage to a motor. Running

capacitors are used in starting winding to increase the running torque of the motor

Starting capacitors are used in the starting winding to increase the starting torque of the motor. Two electrodes or sets of electrodes in the form of plates, separated from each other by an insulating material called the dielectric.

Capacitor Bank

An assembly of capacitors and all necessary accessories, such as switching equipment, protective equipment, controls, etc., required for a complete operating installation.

Capacitor Switching Transients

A one-cycle disturbance that occurs when a substation capacitor bank is initially switched on.

Capacitor Voltage Transformer

A voltage transformer that uses capacitors to obtain a voltage divider effect. It is utilized at EHV voltages instead of an electromagnetic VT for cost and size purposes.

Capacity

A measure of the quantity of instantaneous energy use. The term is applied to the amount of electric power delivered or required for which a generator, turbine, transformer, transmission circuit, station, or system is rated by the manufacturer.

Capacity Factor

The amount of energy that the system produces at a particular site as a percentage of the total amount that it would produce if it operated at rated capacity during the entire year. For example, the capacity factor for a wind farm ranges from 20% to 35%. Thirty-five percent is close to the technology potential.

Capital Recovery Factor (CRF)

A factor used to convert a lump sum value to an annual equivalent.

Captive Customer

A customer who does not have realistic alternatives to buying power from the local utility, even if that customer had the legal right to buy from competitors.

Captive Electrolyte Battery

A battery having an immobilized electrolyte (gelled or absorbed in a material).

Captive Load

Load which may be supplied by an Embedded Generator, in addition to the generator auxiliaries, which is within the Generating Company premises.

Carburizing

The process by which the surface carbon concentration of a ferrous alloy is increased by diffusion from the surrounding environment.

Carnot's Cycle

An ideal reversible four step cycle of operations for the working substance of a heat engine.

Carrier

In a semiconductor, the mobile electrons or holes which carry charges are called carriers.

Carrier Wave

A continuous electromagnetic radiation, of constant amplitude and frequency, emitted by a transmitter. By modulation of the carrier wave, oscillations caused at the transmitting end are conveyed to the receiving end.

Carry Current (in Amps)

The maximum current that can be applied to an already closed contact.

Cartesian Co-Ordinates

The (x,y) co-ordinates of a point in the plane.

Cartridge Fuse

It is a fuse inside a cartridge. The fuse wire is usually enclosed in an evacuated glass tube with metal end caps.

Cartridge Fuse Link

A device comprising a fuse element or several fuse elements connected in parallel enclosed in a cartridge usually filled with arc extinguishing medium and connected to terminations.

Cascade Connection

An arrangement of two or more components or circuits such that the output of one is the input to the next.

Category 5 Cable

Also known as "Cat 5", this cable is used for fast

ethernet and telephone communications. The cable is constructed of 4 twisted pair of copper wire.

Catenary
Curve formed by a chain or string hanging from two fixed points.

Cathode
Is an electrode that, in effect, oxidizes the anode or absorbs the electrons. During discharge, the positive electrode of a voltaic cell is the cathode. When charging, that reverses and the negative electrode of the cell is the cathode.

Cathode Ray Oscilloscope (CRO)
An instrument based upon the cathode ray tube, which provides a visible image of one or more rapidly varying electrical quantities.

Cathode Ray Tube (CRT)
An electron-beam tube in which the beam can be focused to a small cross-section on a luminescent screen and varied in both position and intensity to produce a visible pattern. Electric potentials applied to the deflection plates are used to control the position of the beam, and its movement across the screen, in a desired manner.

Cavitation
Air bubbles in a closed water system. Greatly reduces efficiency in a hydro turbine generator system, and can damage water pumps and pipes in a home water supply system.

Cell
An electrochemical device, composed of positive and negative plates and electrolyte, which is capable of storing electrical energy. It is the basic "building block" of a battery.

Cell Barrier
A very thin region of static electric charge along the interface of the positive and negative layers in a photovoltaic cell. The barrier inhibits the movement of electrons from one layer to the other, so that higher-energy electrons from one side diffuse preferentially through it in one direction, creating a current and thus a voltage across the cell. Also called depletion zone, cell junction, or space charge.

Cell Junction

The area of immediate contact between two layers (positive and negative) of a photovoltaic cell. The junction lies at the center of the cell barrier or depletion zone.

Cell-Reversal (Battery)

Reversing of polarity within a cell od a multi-cell battery due to over discharge.

Centi (c)

Decimal sub-multiple prefix corresponding to one-hundredth or 10^{-2}. This is not a preferred suffix.

Centigrade

Older name for celsius.

Centimeter (cm)

1/100 of a meter, 0.39 inches.

Centimetre-Gram-Second System (Cgs System)

A decimal system which is an earlier form of the metric system.

Central Power

The generation of electricity in large power plants with distribution through a network of transmission lines (grid) for sale to a number of users. Opposite of distributed power.

Centre of Gravity

The centre of gravity of a body is the fixed point through which the resultant force due to the Earth's attraction upon it always passes, irrespective of the position of the body.

Centrifugal Force

The law of centrifugal force is the second law of physics. It states that a spinning object will pull away from its center point and that the faster it spins, the greater the centrifugal force becomes. An example of this would be to tie an object to a string and spin it around, it will try to pull away from you. The faster the object spins, the greater the force that tries to pull the object away. Centrifugal force prevents the electron from falling into the nucleus of the atom. The faster an electron spins, the farther away from the nucleus it will be.

Ceramic

Inorganic, nonmetalllic products for which the interatomic bonding is predominantly ionic.

Cermet

A composite material consisting of a combination of ceramic and metallic materials.

Certificate of Convenience and Necessity

A term used by public service commissions in granting authority to a company to render utility service, usually specifying the area and other conditions of service.

cf/s

A unit of measurement for water flow. Flow equals the volume of water (cubic feet) passing through an area in a given time period (per second). 1 cf/s = 7.48 gallons per second.

Chain Reaction

Any self-sustaining molecular or nuclear reaction, the products of which contribute to the propagation of the reaction.

Channel Associated Signaling (CSA)

Some communications protocols include "signaling" functions along with data. CSA protocols have a method of signaling each data channel (as opposed to a dedicated signaling channel). Also called Robbed Bit Signaling.

Channel-to-channel skew (Ch. To Ch. Skew) (Ps Max)

A signal on one channel has a different phase than the same signal on another channel (delayed/skewed). This is measured in picoseconds, max.

Character (or Comma) Separated Values (CSV)

Character (or Comma) Separated Values Format, format widely utilized for the exchange of data between different software, in which the data are separated by a known character usually a comma.

Characteristic Angle

The angle between the vectors representing two of the energizing quantities applied to a relay used for the declaration of the performance.

Characteristic Curve

A plot or curve displaying the operating values of the characteristic quantities corresponding to various values or combinations of the energizing quantities.

Characteristic Impedance Ratio (CIR)

The maximum value of the

system impedance ratio for which the relay performance remains within the prescribed limits of accuracy.

Charge

The electric charge of an object is a measure of how much electricity is there. It is similar to the mass of an object when you are dealing with gravity, but unlike mass charge can be either positive (+) or negative (-). At the atomic level charge is measured in multiples of the charge on an electron (-1), in larger cases the usual measurement is the Coulomb.

Charge (Battery)

The conversion of electrcal energy from an external source, into chemical energy within a cell or battery.

Charge Carrier

A free and mobile conduction electron or hole in a semiconductor.

Charge Controller

An electronic device which regulates the voltage applied to the battery system from the PV array. Essential for ensuring that batteries obtain maximum state of charge and longest life.

Charge Coupled Device (CCD)

One of the two main types of image sensors used in digital cameras. When a picture is taken, the CCD is struck by light coming through the camera's lens. Each of the thousands or millions of tiny pixels that make up the CCD convert this light into electrons. The accumulated charge at each pixel is measured, then converted to a digital value. This last step occurs outside the CCD, in an analog-to-digital converter (ADC).

Charge Factor

A number representing the time in hours during which a battery can be charged at a constant current without damage to the battery. Usually expressed in relation to the total battery capacity, i.e., C/5 indicates a charge factor of 5 hours. Related to charge rate.

Charge Injection

A parameter pertinent to analog switches. As an analog switch turns on and off, a small amount of charge can be capacitively coupled (injected) from the digital control line to the analog signal path.

Charge Rate

The amount of current applied to battery during the charging process. This rate is commonly expressed as a fraction of the capacity of the battery. For example, the C/2 or C/5.

Charge Rate (Battery)

The rate at which current is applied to a secondary cell or battery to restore its capacity.

Charge Termination Method

Method the battery charger uses to determine when to terminate the charging cycle.

Charged

The condition of a capacitor which has the full charge it can receive from a given applied voltage.

Charge-Retention (Battery)

The tendency of a charges cell or battery to resist self-discharge.

Charging

The process of supplying electrical energy for conversion to stored chemical energy.

Chatter

The undesired intermittent closure of open contacts or opening of closed contacts. It may occur either when the reed contact is operated, released or when subjected to external shock or vibration.

Chemical Reaction

The interaction of two or more substances resulting in chemical changes in them.

Chip

1. Integrated circuit: A semiconductor device that combines multiple transistors and other components and interconnects on a single piece of semiconductor material.
2. Encoding element, in Direct-Sequence Spread Spectrum systems.

Chip Scale Package (CSP)

An IC packaging technology in which solder balls take the place of pins, making the smallest package available. When heated, the solder balls alloy to matching pads on the circuit board.

Chip-Enable Gating

A feature in microprocessor supervisory circuits which prevents the writing of erroneous data when power falls outside of spec. When the main power-supply voltage is below the minimum safe-operating limit, the feature disconnects the chip-enable signal path from the host microprocessor or microcontroller.

Chlorinated Polyethylene

Chlorinated Polyethylene. CPE, a thermoplastic compound, is used to jacket certain types of power cable.

Chlorofluorocarbon

A family of chemicals composed primarily of carbon, hydrogen, chlorine, and fluorine whose principal applications are that of refrigerants and industrial cleansers and whose principal drawback is the tendency to destroy the Earth's protective ozone layer.

Choke

A coil of low resistance and high inductance used in electrical circuits to pass low frequency or direct components while suppressing (or choking) the higher frequency undesirable alternating currents.

Chop

A vertical mode of operation for dual-trace oscilloscopes in which the display is switched between the two channels at some fixed rate. This mode should be used for slow sweep rates.

Chord

The width of a wind turbine blade at a given location along the length.

Chrominance

The color portion portion of a composite video signal. Forms a complete picture once combined with the luminance component.

Circline

A four-contact, double-ended lampholder designed for use with tubular, circular fluorescent lamps.

Circuit

An electrical circuit is the complete path traversed by an electric current.

Circuit (Series)

A circuit supplying energy to a number of devices connected in series. The same current passes through each device in completing its path to the source of supply.

Circuit Breaker

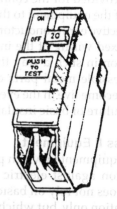

A mechanical switching device capable of making, carrying, and breaking currents under normal conditions. Also making, carrying for a specific time, and automatically breaking currents under specified abnormal circuit conditions, such as those of short circuit. It is usually required to operate infrequently although some types are suitable for frequent operation.

Circuit Extensions

To extend or add-on to an existing circuit to provide an additional power source.

Circuit Insulation Voltage

The highest circuit voltage to earth on which a circuit of a transducer may be used

and which determines its voltage test.

Circuit Protective Conductor (CPC)

A protective conductor connecting exposed conductive parts of equipment to the main earthing terminal.

Circuit Switchers

Circuit-Switchers are multipurpose switching and protection devices. Often used for switching and protection of transformers, single and back-to-back shunt capacitor banks, reactors, lines, and cables. They can close, carry, and interrupt fault currents as well as load currents

Circuit Voltage

The greatest root-mean-square (effective) difference of potential between any two conductors of the circuit.

Circular Mil

A unit of area used in measuring the cross section of fine wire. 1 circular mil corresponds to the area of a circle whose diameter is equal to one-thousandth of an inch. 1 circular mil = $0.5067 \cdot 10^{-9} \text{ m}^2$

Circularly Polarised Light

Light which can be resolved into two vibrations lying in planes at right angles, of equal amplitude and frequency and differing in phase by 90°.

Circular-Mil (cmil)

The area of a circle with a diameter of one mil (1/1000 inch), used to describe the cross-sectional area of a conductor. One cmil equals approximately 0.0000008 square inches.

Clamping Level

The voltage point at which a surge protector begins to limit surges.

Clamping Voltage

The peak voltage that can be measured after a Surge Protective Device has limited or "clamped" a transient voltage surge. Clamping voltage must be determined by using IEEE Standard C62 testing and evaluated by UL Standard 1449.

Class I Equipment

Equipment in which protection against electric shock does not rely on basic insulation only, but which includes an additional safety precaution in that means are provided for the connection of the equipment to the protective earth conductor in the fixed wiring of the installation in such a way that accessible metal parts cannot become live in the event of a failure of basic insulation.

Class II Equipment

Equipment in which protection against electric shock does not rely on basic insulation only, but which additional safety precautions such as double insulation or reinforced insulation are provided, there being no provision for protective earthing or reliance upon installation conditions.

Class III Equipment

Equipment in which protection against electric shock relies on supply at SELV and in which voltages higher than those of SELV are not generated.

Class Index

A number which designates an accuracy class.

Clean Power

Electrical power which has been conditioned and/or regulated to remove electrical noise from the output power.

Clearing Time

The total time needed for a protective device such as a fuse or circuit breaker to clear a fault.

Cleavage of Lateral Epitaxial Films for Transfer (CLEFT)

A process for making inexpensive GaAs photovoltaic cells in which a thin film of GaAs is grown atop a thick, single-crystal GaAs (or other suitable material) substrate and then is cleaved from the substrate and incorporated into a cell, allowing the substrate to be reused to grow more thin-film GaAs.

Click/Pop Reduction

A feature that eliminates "clicks" and "pops" —unwanted transient noise signals during power-up, shutdown, connection, etc.

Client

As part of a computer network, where a server is employed, this is the customer or non-server side. When you log onto a server, from another computer, the word "Client" refers to you, your computer or your software.

Climbers

Hooks for climbing poles that are attached to a lineman's boots.

Clock

The basic timing signal in a digital system.

Clock and Data Recovery

The process of extracting and reconstructing clock and data information from a single-wire/channel, serial data stream.

Clock Hanger

A single, recessed receptacle with a specialized cover plate that provides a hook or other means of supporting a wall clock.

Clock Jitter

A periodic waveform (especially a clock) is expected to cross certain thresholds at precisely timed moments. Variations from this ideal are called jitter.

Clock Throttling

Reducing the frequency or duty-cycling the clock of an integrated circuit usually for the purpose of reducing heat generation.

Close Circuit
A circuit permitting a continuous current.

Close differential
A differential of minimum 70%

Closed Entry Contact
A female contact designed to prevent the entry of a pin or probing device having a cross-sectional dimension (diameter) greater than the mating pin.

Closing Impulse Time
The time during which a closing impulse is given to the circuit breaker.

Closing Time
Referring to a circuit breaker it is the necessary time for it to close, beginning with the time of energizing of the closing circuit until contact is made in the CB.

Cloud Enhancement
The increase in solar intensity caused by reflected irradiance from nearby clouds.

Coaxial Cable
Also known as "Coax", this cable is typically used to connect TV to its video source. Coaxial Cable consists of a small copper wire or tube, surronded by an insulating material and another conductor with a larger diameter, normally copper braid or a conductive tube. The cable is jacketed for mechanical and electrical protection.

Code Corrections
Procedure designed to eliminate wiring conditions that do not meet National Electrical Code requirements and safety conditions.

Code Division Multiple Access (CDMA)
CMDA is a technique used mainly with personal communications devices such as mobile phones that digitizes the conversation and tags it with a special frequency code. The data is then scattered across the frequency band. The receiving device is instructed to decipher only the data corresponding to a particular code to reconstruct the signal.

Code Letter
A letter which appears on the nameplates of AC motors to show their locked-rotor kilovolt-amperes per

horsepower at rated voltage and frequency.

Code of Conduct

A series of requirements applicable to electric utilities and alternate electric suppliers engaged in retail marketing and other transactions, as approved by the Commission in case No. U-12134. The Commission's Code of Conduct promotes fair competition by establishing measures to prevent cross-subsidization, information sharing, and preferential treatment.

Coefficient

A number or other known factor or multiplier which measures some specified property of a given substance or algebraic expression.

Coefficient of coupling (k)

A numerical rating between 0 and 1 that specifies the degree of magnetic coupling between two circuits. Maximum coupling is 1 and no coupling is 0.

Coefficient of Utilization (CU)

The percentage of light generated within a luninaire which ultimately strikes the work surface. It is usually

expressed as a decimal percentage.

Coercive Force

The strength of the magnetic field to which a ferromagnetic material undergoing a hysteresis cycle must be subjected in order to demagnetise the material completely.

Coercivity

Demagnetization force required to reduce polarization or induction to zero.

Coffin Hoist

A chain hoist of any type.

Cogeneration

Production of heat energy and electrical or mechanical power from the same fuel in the same facility. A typical cogeneration facility produces electricity and steam for industrial process use.

Coil

An assemblage of successive convolutions of a conductor. A unit of a winding consisting of one or more insulated conductors connected in series and surrounded by common insulation, and arranged to link or produce magnetic flux.

Coil Loss

Power loss in a transformer due to the flow of current. These losses are present only when the transformer is serving a load. Load losses vary by the square of the current magnitude. Load losses are composed of losses due to the current flow through the resistance of the conductors as well as eddy losses in the windings and stray losses due to current flow through other components.

Coincidence Factor

The ratio of the coincident maximum demand of two or more loads to the sum of their noncoincident maximum demands for a given period. The coincidence factor is the reciprocal of the diversity factor and is always less than or equal to one.

Coincidental Peak Load

The sum of two or more peak loads that occur in the same time interval.

Cold

Refers to non-energized equipment, lines or circuits.

Cold Cathode Fluorescent

Lighting (CCFL)

Often used as a backlight for LCD displays.

Cold Cathode Fluorescent Tube (CCFT)

Often used as a backlight for LCD displays.

Cold Cathode Lamp

An electric-discharge lamp whose mode of operation is that of a glow discharge (Neon Lights).

Cold-Weather Ballast

Compact fluorescent light bulbs require a ballast to regulate the voltage of the electricity that is applied to the gas inside the lamp. Below-freezing weather can adversely affect the electronic components in these ballasts, causing most compact fluorescent bulbs to appear dim in cold weather. Cold-weather ballasts compensate for this problem and keep the bulb glowing brightly, even in weather as cold as -10°F (-23°C).

Collector

The electrode in a transistor through which a primary flow of carriers leaves the inter-electrode region. It is called a collected because it collects or gathers the carriers sent into the base by the emitter.

Collector Roadway (Lighting)

The distributor and collector roadways servicing traffic between major and local roadways. These are roadways used mainly for traffic movements within residential, commercial and industrial areas.

Colour

The sensation of colour is the result of the interpretation by the human central nervous system of the effect produced upon the eye by electromagnetic radiation of a particular wave length.

Colour Temperature

The temperature of a full radiator (radiation of all frequencies) which would emit visible radiation of the same spectral distribution as the radiation from the light source under consideration. [Unit kelvin, K]

Combination

A multiple gang wallplate with openings in each gang to accommodate different devices.

Combination Unilay

A stranding comfiguration that uses two strand sizes to achieve a 3% reduction in the conductor diameter without conpression

Combined Collector

A photovoltaic device or module that provides useful heat energy in addition to electricity.

Combined Cycle

An electric generating technology in which electricity is produced from otherwise lost waste heat exiting from one more gas (combustion) turbines. The exiting heat is routed to a conventional boiler or to a heat recovery steam generator for utilization by a steam turbine in the production of electricity. This process increases the efficiency of the electric generating unit.

Combined Cycle Plant

An electric generating station that uses waste heat from its gas turbines to produce steam for conventional steam turbines.

Combined Cycle Unit

An electric generating unit that consists of one or more

combustion turbines and one or more boilers with a portion of the required energy input to the boiler(s) provided by the exhaust gas of the combustion turbine(s).

Combined Heat and Power (CHP)

A plant that generates electricity and supplies thermal energy, typically steam, to an industrial or other heating requirement.

Combined Pumped-Storage Plant

A pumped-storage hydroelectric power plant that uses both pumped water and natural streamflow to produce electricity.

Combustion or Burning

A chemical reaction or complex chemical reaction, in which a substance combines with oxygen producing heat, light and flame.

Combustion Turbine

A type of generating unit normally fired by oil or natural gas. The combustion of the fuel produces expanding gases, which are forced through a turbine, thereby generating electricity.

Come-a-long

A wire grip for holding a conductor or strand under tension.

Commercial

The commercial sector is generally defined as nonmanufacturing business establishments, including hotels, motels, restaurants, wholesale businesses, retail stores, and health, social, and educational institutions. The utility may classify commercial service as all consumers whose demand or annual use exceeds some specified limit. The limit may be set by the utility based on the rate schedule of the utility.

Commercial (Lighting)

A business area of a municipality where there are ordinarily many pedestrians during night hours. The definition applies to densely developed business area outside, as well as within, the central part of a municipality. The area contains land use attracting a relatively heavy volume of nighttime vehicular and/or pedestrian traffic on a frequent basis.

Commercial Operation

Commercial operation begins when control of the loading of the generator is turned over to the system dispatcher.

Commercialization

Programs or activities that increase the value or decrease the cost of integrating new products or services into the electric sector.

Common

The potential level serving as ground for the entire circuit.

Common Mode (CM)

The term refers to electrical interference which is measurable as a ground referenced signal. In true common mode a signal is common to both the current carrying conductors.

Common Mode Noise

Electrical noise or interference between each of the conductors and ground

Common Mode Signal

A signal that is applied with equal strength to both inputs of a differential amplifier or an operational amplifier.

Common Mode Tripping

Automatic removal of two or more generating plant from the system owing to a cause that is common to both or all the generators.

Compact Fluorescent Lamp (CFL)

A fluorescent lamp in compact form that may be conveniently used, in normal holders, in place of the lesser efficient incandescent lamps. The lamp life is significantly longer than incandescent lamps.

Compact Stranding

A stranding configuration with concentric strands in which each layer is passed through a compacting die to reduce the conductor diameter by approximately 10%

Compander

Signal processing technique which uses both compression and expansion to improve dynamic range and signal-to-noise ratio.

A signal is passed through a non-linear transformation prior to transmission. A reverse of this transformation occurs at reception. The transformation is such that quiet portions are boosted and loud portions reduced. Noise is reduced because the quiet signals are louder, compared to the noise in the transmission channel.

Compass

The magnetic compass is used to obtain the direction of the earth's magnetic field at a point. In its simplest form consists of a magnetized needle pivoted at its centre so that it is free to move in a horizontal plane.

Complex Number

Consists of two parts, real and imaginary. They obey the ordinary laws of algebra except that their real and imaginary parts must be equated separately.

Compliance Voltage

The specified maximum voltage that a transducer (or other device) current output must be able to supply while maintaining a specified accuracy.

Component Lead

The solid or stranded wire or formed conductor that extends from a component and serves as a readily formable mechanical or electrical connection or both.

Composite

A material brought about by combining materials differing in composition or form on a macroscale for the purpose of obtaining specific characteristics and properties. The constituents retain their identity such that they can be physically identified and they exhibit an interface between one another.

Compound

An insulating or jacketing material made by mixing two or more ingredients

Compound Motor

A d.c. motor with both a series connected winding as well as a shunt connected winding. Depending on whether the fields of the series winding and the shunt winding aid each other or oppose each other, they are called cumulative compound or differential compound.

Compressed

A stranding configuration with concentric strands in which either all layers or the outer layer only is passed through a die to reduce the conductor diameter by 3%

Compressed-Air Energy Storage (CAES)

CAES plants use off-peak electrical energy to compress air into underground storage reservoirs for storage until times of peak or intermediate electricity demand. Wind power offers a good opportunity for charging CAES storage. The storage is typically underground in natural aquifers, depleted oil or gas fields, mined salt caverns, or excavated or natural rock caverns. To generate power, the compressed air is first heated by gas burners, then passed through a turbine.

Compression Splice

A compression connector used to join two conductors. There are different designs used for overhead and underground conductors. For overhead conductors, there are different designs for limited and full tension applications.

Concentrator

A PV module that uses optical elements to increase the amount of sunlight incident on a PV cell. Concentrating arrays must track the sun and use only the direct sunlight because the diffuse portion cannot be focused onto the PV cells. Efficiency is increased, but lifespan is usually decreased due to the high heat.

Concentrator Array

A PV array which uses concentrating devices (reflectors, lenses) to increase the intensity of the sunlight striking the array.

Concentric

Circles having the same centre.

Concentric Stranding

A stranding configuration in which individual wires are stranded concentrically with no reduction in overall diameter. Typically used for bare conductors

Concentricity

In a wire or cable, the measurement of the location of the center of the conductor with respect to the geometric center of the surrounding insulation.

Concrete

A composite material consisting of aggregate particles bound together in a solid body by a cement.

Condensation Polymerization

The formation of polymers by an intermolecular reaction involving at least two monomer species, usually with the production of a low molecular weight by-product such as water.

Conduct

The ability of two conductors separated by a dielectric to store electricity when a potential difference exists between the conductors.

Conductance

The conductance is the reciprocal of the resistance in a resistive circuit, or the real part of the admittance in a complex circuit. It is also the ability of an element to conduct electric current [Unit: siemens or S]

Conduction Band (or conduction level)

An energy band in a semiconductor in which electrons can move freely in a solid, producing a net transport of charge.

Conductivity

The ability of a material to conduct electric current. It is expressed in terms of the current per unit of applied voltage. It is the reciprocal of resistivity.

Conductor

An electrical path which offers comparatively little resistance. A wire or combination of wires not insulated from one another, suitable for carrying a single electric current. Bus bars are also conductors. Conductors may be classed with respect to their conducting power as; (a) good; silver, copper, aluminum, zinc, brass, platinum, iron, nickel, tin, lead; (b) fair; charcoal and coke, carbon, plumb ago, acid solutions, sea water, saline solutions, metallic ores, living vegetable substances, moist earth; (c) partial; water, the body, flame, linen, cotton, mahogany, pine, rosewood, lignum vitae, teak, and marble.

Conductor loss

Loss occurring in a conductor due to the flow of current. Also known as the I^2R loss and copper loss.

Conductor Shield

A semiconducting material, normally cross-linked polyethyene, applied over the conductor to provide a smooth and compatible interface between the conductor and insulation. This smooth semiconducting shield is at the same potential as the conductor resulting in dielectric field lines that are not distorted.

Conduit
A tubular raceway for data or power cables. Metallic conduit is common, although non-metallic forms may also be used. A conduit may also be a path or duct and need to be tubular.

Conduit Fill
Volumetric measurement of the duct space occupied by the cables inside, expressed as a percent.

Congestion
A condition that occurs when insufficient transfer capacity is available to implement all of the preferred schedules for electricity transmission simultaneously.

Congruent Transformation
A transformation of one phase to another that does not involve any change in composition.

Conjunctive Test
A parametric or specific test of a protection system on all components and auxiliary equipment that are connected.

Connection
The physical connection (e.g. transmission lines, transformers, switch gear, etc.) between two electric systems permitting the transfer of electric energy in one or both directions.

Connector
A device providing electrical connection/disconnections. It consists of a mating plug and receptacle. Various types of connectors include DIP, card edge, two-piece, hermaphroditic and wire-wrapping configurations. Multiple contact connectors join two or more conductors with others in one mechanical assembly.

Connector Discontinuity
An ohmic change in contact resistance.

Connector Insert
For connectors with metal shells, the insert holds contacts in proper arrangement while electrically insulating

them from each other and from the shell.

Connector Shell

The case that encloses the connector insert and contact assembly. Shells of mating connectors can protect projecting contacts and provide proper alignment.

Conservation

Reducing energy consumption and energy waste using a strategy to attain higher efficiency in energy production and utilization, to accommodate behavior to maximize personal welfare in response to changing prices, and shifting from scarce to more plentiful energy resources.

Conservation and Other Demand-Side Management (DSM)

This Demand-Side Management category represents the amount of consumer peak load reduction at the time of system peak due to utility programs that reduce consumer load during many hours of the year. Examples include utility rebate and shared savings activities for the installation of energy efficient appliances, lighting and electrical machinery, and weatherization materials. In addition, this category includes all other Demand-Side Management activities, such as thermal storage, time-of-use rates, fuel substitutions, measurement and evaluation, and any other utility-administered Demand-Side Management activity designed to reduce demand and/or electricity use.

Conservation of Mass

The law of conservation of mass states that in any system matter cannot be created or destroyed.

Conservation of Mass and Energy

A principle resulting from Einstein's special theory of relativity, which combines the separate laws of conservation of mass and energy. It states that in any system the sum of the mass and energy remains constant.

Conservation of Momentum

The law of conservation of momentum states that for a perfectly elastic collision, the total momentum of two bodies before impact is equal to their total momentum of two bodies before impact is equal to their total momentum after impact.

Constant Current Charge

Charging technique where the output current of the charge source is held cunstant.

Constant Horsepower Motor

A term used to describe a multi-speed motor in which the rated horsepower is the same for all operating speeds. When applied to a solid state drive unit, it refers to the ability to deliver constant horsepower over a predetermined speed range.

Constant Potential Charge

Charging technique where the output voltage of the charge source is held constant and the current is limited only by the resistance of the battery.

Constant Torque Motor

A multi-speed motor for which the rated horsepower varies in direct ratio to the synchronous speeds.

Constant-Current Charge

A charging process in which the current applied to the battery is maintained at a constant value.

Constant-Speed Wind Turbines

Turbines that operate at a constant rotor revolutions per minute (RPM) and are optimized for energy capture at a given rotor diameter at a particular speed in the wind power curve.

Constant-Voltage Charge

A charging process in which the voltage applied to a battery is held at a constant value.

Construction Work In Progress (CWIP)

The balance shown on a utility's balance sheet for construction work not yet completed but in process. This balance line item may or may not be included in the rate base.

Consumer Unit

(may also be known as a consumer control unit or electricity control unit) A particular type of distribution board comprising a co-ordinated assembly for the control and distribution of electrical energy, principally in domestic premises, incorporating manual means of double pole isolation on the incoming circuit(s) and an assembly of one or more fuses, circuit breakers, residual current operated devices or signalling and other devices purposely manufactured for such use.

Consumption (Fuel)

The amount of fuel used for gross generation, providing standby service, start-up and/or flame stabilization.

Contact

The current-carrying parts of a reed contact that engage or disengage to close or open electrical circuits.

Contact Discharge

An ESD test method where the ESD generator makes direct contact with the device under test (DUT).

Contact Gap

The distance between mating reed contacts when the contacts are open.

Contact Plating

Plated-on metal applied to the base contact metal to provide the required contact resistance and/or wear resistance.

Contact Rating (in Watts)

The maximum power, a reed contact can switch.

Contact Resistance

Maximum permitted electrical resistance of pin and socket contacts when assembled in a connector under typical service use.

Contact Retainer

A device either on the contact or in the insert to retain the contact.

Contact Size

Defines the largest size wire that can be used with the specific contact. By specification dimensioning, it also defines the diameter of the engagement end of the pin.

Contact, Female

A contact located in an insert or body in such a manner that the mating contact is inserted into the unit. It is similar in function to a socket contact.

Contact, Male

A contact located in an insert or body in such a manner that the mating portion extends into the female contact. It is similar in function to a pin contact.

Contactor

An electro-mechanical device that is operated by an electric coil and allows automatic or remote operation to repeatedly establish or interrupt an electrical power circuit. A contactor provides no overload protection as required for motor loads.

Continuity

The state of being whole, unbroken.

Continuity Test

A test performed on a conductor to determine if it is unbroken throughout its length

Continuous Load

A load where the maximum current is expected to continue for three hours or more. Rating of the branch circuit protection device shall not be less tan 125% of the continuous load.

Continuous Rating

The constant voltage or current that a device is capable of sustaining. This is a design parameter of the device.

Contract Path

The most direct physical transmission tie between two interconnected entities. When utility systems interchange power, the transfer is presumed to take place across the "contract path", notwithstanding the electric fact that power flow in the network will distribute in accordance with network flow conditions. This term can also mean to arrange for power transfer between systems.

Contract Price

Price of fuels marketed on a contract basis covering a period of 1 or more years. Contract prices reflect market conditions at the time the contract was negotiated and therefore remain constant throughout the life of the contract or are adjusted through escalation clauses. Generally, contract prices do not fluctuate widely.

Control System

A circuit that controls the operation of a device, such as the start/stop circuit on a motor starter.

Controller

A device or group of devices that serves to govern, in some predetermined manner, the electric power delivered to the apparatus to which it is connected.

Controller Area Network (CAN)

The CAN protocol is an international standard defined by ISO-11898.

Convection

Transference of heat through a liquid or gas by the actual movement of the fluid.

Conversion Efficiency (Cell Or Module)

The ratio of the electric energy produced by a photovoltaic device (under one-sun conditions) to the energy from sunlight incident upon the cell.

Converter

Any technology that changes the potential energy in fuel into a different form of energy such as heat or motion. The term also is used to mean an apparatus that changes the quantity or quality of electric energy.

Convolution

The convolution of two signals consists of time-reversing one of the signals, shift-ing it, and multiplying it point by point with the second signal, and integrating the product. It is used to characterize physical systems.

Cooling System

Energy Efficiency program promotion aimed at improving the efficiency of the cooling delivery system, including replacement, in the residential, commercial, or industrial sectors.

Coordination

Relating to the protection of the power system, the process of coordinating the fuse, breakers and reclosers of a system so to allow the down stream devices to operate first.

Coordination Number

The number of atomic or ionic nearest neighbors.

Coplanar Line

A line which is in the same plane as another line. Any two intersecting lines must lie in the same plane, and therefore be coplanar.

Copolymer

A polymer that consists of two or more dissimilar mer

units in combination along its molecular chains.

Copper Indium Diselenide (Cuinse₂, Or CIS)

A polycrystalline thin-film photovoltaic material (sometimes incorporating gallium (CIGS) and/or sulfur).

Copper Loss

Same as conductor loss. Conductors were traditionally made of copper, hence the name.

Cord Connector

A portable receptacle designed for attachment to or provided with flexible cord, not intended for fixed mounting.

Core Balance Current Transformer

A ring-type current transformer in which all primary conductors are passed through the aperture making any secondary current proportional to any imbalance in current.

Core Loss

Power loss in a transformer due to excitation of the magnetic circuit (core). No load losses are present at all times when the transformer has voltage applied. No load losses vary based on the applied voltage, and are essentially constant whether the transformer is supplying a load or not.

Corona Discharge

An electrical discharge at the surface of a conductor accompanied by the ionization of the surronding atmosphere. It is normally accompanied by light and audible noise.

Corona Inception

Inception of the ionisation of the air on the surface of a conductor, caused by the voltage gradient exceeding a critical value, but not being sufficient to cause sparking or flash over.

Corona Loss

Power loss due to corona.

Corrosion

Deteriorative loss of a metal as a result of dissolution environmental reactions.

Corrosion Resistant

An inlet constructed of special materials and/or suitably plated metal parts that is designed to withstand corrosive environments. Corrosion resistant devices must pass the ASTM B117-13 five-hundred hour Salt Spray (Fog) Test with no visible corrosion.

Cost

The amount paid to acquire resources, such as plant and equipment, fuel, or labor services.

Cost-of-Service Regulation

Traditional electric utility regulation under which a utility is allowed to set rates based on the cost of providing service to customers and the right to earn a limited profit.

Co-Tree of Network

Complement of the tree of the network.

Coulomb (c)

A unit of electric charge in SI units (International System of Units). A Coulomb is the quantity of electric charge that passes any crossection of a conductor in one second when the current is maintained constant at one ampere.

Coulomb's Law

The force of attraction or repulsion between two point charges is proportional to the magnitude of the charges and inversely proportional to the square of the distance between them.

Counter Electromotive Force (CEMF)

The voltage induced in a wire by self-induction which opposes the applied voltage. Also called back emf.

Counting Relay

A relay that counts the number of times it is energized and actuates an output after a desired count has been reached.

Coupling

The association of two or more circuits or systems in such a way that power or information may be transferred from one to the other.

Coupling Capacitor

A capacitor used to transmit an ac signal from one node to another.

Coupling Coefficient

The coupling coefficient of a pair of coils is a measure of the magnetic coupling between two coils.

Covalent Bond
A primary interatomic bond that is formed by the sharing electrons between neighboring atoms.

Cramming
The practice of adding charges to a customer's monthly bill for services that the customer has not authorized. This is an illegal practice.

Creep
The time-dependent permanent deformation that occurs under stress; for most materials it is important only at elevated temperatures.

Creepage Distance
The shortest distance between two conductors as measured along the device that separates them. Creepage Distance is normally a design parameter of insulators or insulating bushings.

Crest Factor
The ratio of the peak or maximum value of a wave, to the r.m.s. value.

Crest Value
The maximum value of a wave form. This is normally associated with electrical fault magnitude or transients.

Critical Damping
Critical damping of a measuring instrument causes the equilibrium deflection to be reached in the shortest possible time, with the oscillations of the needle being quickly damped out.

Critical Mass
The minimum amount of fissile material required in a nuclear reactor to sustain a chain reaction.

Cross
Any accidental contact between electric wires or conductors.

Cross-Bonding
A method of connecting the sheaths of single core cables in a three phase system in order to reduce the circulating currents flowing in the sheaths.

Cross-Linked Polyethylene (XLPE)
A Common thermoset insulation material for building wire and cable Polyethylene made from petroleum and natural gas. It undergoes a

crosslinking chemical reaction during a curing process that causes the compound molecules to bond, forming heavier molecules with desired physical and chemical properties.

Crosslinked Polymer

A polymer in which adjacent linear molecular chains are joined at various positions by covalent bonds.

Crosstalk

"Leakage" of signal from one source into another.

Crowbar

1. Crowbar circuit: A power supply protection circuit that rapidly short-circuits ("crowbars") the supply line if the voltage and/or current exceeds defined limits. In practice, the resulting short blows a fuse or triggers other protection, effectively shutting down the supply. Usually achieved by an SCR or other silicon device, or by a mechanical shorting device.

Probably named for the concept of using a big metal bar to mechanically provide the short circuit, as might be used done in a high-current application; or from the appearance of a crowbar circuit's I-V curve.

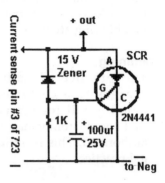

2. Shoot-through current: In a push-pull amplifier stage, one transistor pushes current to the output, toward a positive voltage, and a second device pulls down. These are designed so both devices are never fully on, which would effectively short the power supply. Events that allow both devices to be on (e.g. circuit faults or a brief moment in the switching cycle) are said to "crowbar" the circuit.

Cryptanalysis

The art and science of breaking encryption or any form of cryptography.

Cube Tap

An adapter that converts one receptacle opening into multiple openings.

Cumulo-Nimbus Cloud

Thunder cloud from which lightning strikes originate.

Curie

Measure of the activity of a radioactive substance. It is defined in terms of the rate of decay of a quantity of a radioactive isotope.

Curie Temperature

That temperatue above which a ferromagnetic or ferrimagnetic material becomes paramagnetic.

Current

The electrical current is simply a measure of how much electricity passes a given point in a fixed amount of time. It is similar to the current of a stream or river, which measures how much water passes a given point in a fixed amount of time. Electrical current is measured in Amperes.

Current at Maximum Power

The current at which maximum power is available from a module.

Current Balance

Instrument for the determination of an electric current in absolute electromagnetic units.

Current Carrying Capacity of a Conductor

The maximum current which can be carried by a conductor under specified conditions without its steady state temperature exceeding a specified value.

Current Limiting Fuse

A fuse designed to operate at the current zero crossing.

Current Mode Feedback

An alternative op amp topology usually used in high-speed amplifiers. It is sensitive to feedback impedance, and cannot be used as an integrator.

Current Rating

This is the maximum current, which the fuse will carry for an indefinite period without undue deterioration of the fuse element.

Current Source

A source which essentially maintains the source current at a predefined value almost independent of the load conditions.

Current Squared Times Time (I^2t)

This is an electrical quantity that is used to determine

energy to a protective device, such as ·a circuit breaker or fuse.

Current Tap

An adapter designed for medium base lampholders which has one or two receptacle openings. Available with or without integral switch.

Current Transformer

An instrument transformer specifically designed to give an accurate current ratio for measurement and/or control purposes. They are always connected in series with the circuit (like an ammeter) and hence should never be allowed to have their secondary to be on open circuit to avoid saturation

Current/Voltage Curve (I-V curve)

A current/voltage curve, which expresses the possible combinations of current and voltage output of a photovoltaic device.

Current-Mode Controller

A DC-DC switching regulator which regulates its output voltage by varying the peak inductor current on a cycle-by-cycle basis to output a regulated voltage despite variations in load-current and input-voltage.

Current-Sense Amplifier

An amplifier that measures current by measuring the voltage drop across a resistor placed in the current path. The current sense amp outputs either a voltage or a current that is proportional to the current through the measured path.

Cut Off Voltage

Battery Voltage reached at the termination of a discharge. Also Known as the End Point Voltage (EPV).

Cut-In

The rotational speed at which an alternator or generator starts pushing electricity hard enough (has a high enough voltage) to make electricity flow in a circuit.

Cutoff (Lighting)

Luminaire light distribution is classified as cutoff when the candlepower per 1000 lamp lumens does not numerically exceed 25 (2.5%) at an angle of 90 degrees above nadir (horizontal), and 100 (10%) at a vertical angle of 80 degrees above nadir. This applies to any lateral angle around the luminaire.

Cycle

A complete reversal of alternating current, passing through a complete set of changes or motions in opposite directions, from a rise to maximum, return to zero, rise to maximum in the other direction, and another return to zero. One complete positive and one complete negative alternation of current or voltage.

Cycle (Battery)

A sequence of discharge followed by a charge, or a charge folowed by a discharge, of a battery under specific conditions.

Cycle Life

For rechargeable batteries, the total number of charge/discharge cycles the cell can sustain before it's capacity is significantly reduced. End of life is usually considered to be reached when the cell or battery delivers only 80% of rated ampere- hour capacity. NiMH batteries typically have a cycle life of 500 cycles, NiCd batteries can have a cycle life of over 1,000 cycles. The cycle of a battery is greatly influenced by the type depth of the cycle (deep or shallow) and the method of recharging. Improper charge cycle cutoff can greatly reduce the cycle life of a battery.

Cyclic Redundancy Check (CRC)

A check value calculated from the data, to catch most transmission errors. A decoder calculates the CRC for the received data and compares it to the CRC that the encoder calculated, which is appended to the data. A mismatch indicates that the data was corrupted in transit. Depending on the algorithm and number of CRC bits, come CRCs contain enough redundant information that they can be used to correct the data.

Czochralski Process

A method of growing large size, high quality semiconductor crystal by slowly lifting a seed crystal from a molten bath of the material under careful cooling conditions.

Daily Peak

The maximum amount of energy or service demanded in one day from a company or utility service.

Daisy Chain

A method of propagating signals along a bus in which the devices are connected in series and the signal passed from one device to the next. The daisy chain scheme permits assignment of device priorities based on the electrical position of the device on the bus.

Dallastat

Dallas Semiconductor trademark: Refers to any of the Dallas family of digital rheostats.

Damping

Decrease in the amplitude of an oscillation or wave motion with time.

Dangling Bonds

A chemical bond associated with an atom on the surface layer of a crystal. The bond does not join with another atom of the crystal, but extends in the direction of exterior of the surface.

Data

Information in numeric, alphanumeric, or other form.

Data Processing Equipment

Electrically operated machine units which, separately or assembled in systems, accumulate, process and store data. Acceptance and divulgence of data may or may not be by electronic means.

dBm

A unit that defines a signal level by comparing it to a reference level. The reference level of 0dBm is defined as 1mW. The signal level in dBm is 10 times the log of the signal's power over that of the 0 dBm reference.

DC to DC Converter

Electronic circuit to convert dc voltages (e.g., PV module voltage) into other levels (e.g., load voltage). Can be part of a maximum power point tracker (MPPT).

DC-DC Controllers

A DC-DC converter (switch-mode power supply) in which the power switch (usually a power MOSFET) is external to the IC.

Dead

Free from any electric connection to a source of potential difference and from electric charge. The term is used only with reference to current carrying parts that are sometimes alive.

Dead Front

Generally refers to equipment that is connected without exposed conductor. Dead front equipment is normally connected with elbows. Thomas & Betts Elastimold division manu-factures elbows and other medium voltage connection devices.

Dead Time

The time between the fault arc being extinguished and the circuit breaker contacts re-mating.

Debounce

Electrical contacts in mechanical pushbutton switches often make and break contact several times when the button is first pushed. A debouncing circuit removes the resulting ripple signal, and provides a clean transition at its output.

Deci (d)

Decimal sub-multiple prefix corresponding to one-tenth or 10^{-1}. This is not a preferred suffix.

Decibels (dB)

A method for specifying the ratio of two signals. dB=10 times the log of the ratio of the power of the two signals. This is equal to 20 times the ratio of their voltages, if the signals are driving equal impedances. Decibels are also used to describe a signal level by comparing it to a reference level. The reference is usually defined as 0dB and the dB value of the signal is 10 times the log of the signal's power over that of the reference. A letter is sometimes added to signify the reference. For instance, dBm is relative to 0 dBm=1mW.

De-energized
Free from any electrical connection to a source of potential difference and from electrical charge.

Deep Cycle
A cycle in which the discharge is continued until the battery reaches it's cut-off voltage, usually 80% of discharge.

Deep Cycle battery
A battery designed to regularly discharge 80 percent of its capacity before recharging.

Deep Discharge
Discharging a battery to 20-percent or less of its full charge.

Deep-Cycle Battery
A battery with large plates that can withstand many discharges to a low state-of-charge.

Defibrillation
The use of electric shock to stop abnormally fast heart rhythms. Electrical current is used to restore the heart's natural pacemaker function which resumes a normal heartbeat. The shock is administered through electrodes placed on the chest wall (external defibrillation) or in the heart (internal defibrillation).

De-ionization Time
The time required for dispersion of ionized air after a fault is cleared so that the arc will not re-strike on re-energization.

Deka (da)
Decimal multiple prefix corresponding to ten or 10. This is not a preferred suffix.

Delay Angle or Control Angle
The control angle for rectification (also known as the ignition angle) is the angle by which firing is delayed beyond the natural take over for the next thyristor.

Delivering Party
The entity supplying the capacity and/or energy to be transmitted at Point(s) of Receipt.

Delta Connection

A method of connecting three elements of a three-phase electrical system in a closed triangle or delta, and with the three phases being taken from the corners of the triangle.

Delta-Wye

Refers to a transformer that is connected Delta on the primary side and Wye on the secondary.

Demagnetization

The process of removing the magnetic properties from a material.

Demilitarized Zone (DMZ)

Networking has corrupted the term and used it to refer to an unprotected subnet connected to a local network, but outside the protection of a firewall.

Demodulation

The process or separating information from a modulated carrier wave.

Demonstration

The application and integration of a new product or service into an existing or new system. Most commonly, demonstration involves the construction and operation of a new electric technology interconnected with the electric utility system to demonstrate how it interacts with the system. This includes the impacts the technology may have on the system and the impacts that the larger utility system may have on the functioning of the technology.

Dendrite

A slender threadlike spike of pure crystalline material, such as silicon.

Dendritic Web Technique

A method for making sheets of polycrystalline silicon in which silicon dendrites are slowly withdrawn from a melt of silicon whereupon a web of silicon forms between the dendrites and solidifies as it rises from the melt and cools.

Dense Wave Division Multiplexing (DWDM)

The technology by which the frequencies of light carried on a single optical fiber are subdivided into discrete wavelengths, allowing for the greater transmission of data.

Density
The mass of unit volume of a substance. [Unit: kg/m³].

Dependable Capacity
The system's ability to carry the electric power for the time interval and period specified. Dependable capacity is determined by such factors as capability, operating power factor and portion of the load the station is to supply.

Dependent Time Measuring Relay
A measuring relay for which times depend, in a specified manner, on the value of the characteristic quantity.

Depletable Energy Sources
This includes:
1) electricity purchased from a public utility and
2) energy obtained from burning coal, oil, natural gas or liquefied petroleum gasses.

Depletion Zone
Same as cell barrier. The term derives from the fact that this microscopically thin region is depleted of charge carriers (free electrons and holes).

Depreciation, Straight-line
Straight-line depreciation takes the cost of the asset less the estimated salvage value and allocates the cost in equal amounts over the asset's estimated useful life.

Depth of Discharge
The amount of energy that has been removed from a battery (or battery pack). Usually expressed as a percentage of the total capacity of the battery. For example, 50% depth of discharge means that half of the energy in the battery has been used. 80% DOD means that eighty percent of the energy has been discharged, so the battery now holds only 20% of its full charge.

Derating
Calculations that reduce standard tabulated ratings based, generally based on ambient temperature or proximity to a heat source.

Data Encryption Standard (DES)
DES is an encryption, method that uses an algorithm for private key encryption, in which the sender uses the same private key as the recipient uses to decode it.

Derating Factor

A value that tells how much to reduce the power rating of a device for each degree above the reference temperature.

Deregulation

The process of changing the laws and regulations that control the electric industry to allow competition and customer choice of energy supply. This is also known as Restructuring.

Derived Units

Units of physical measurement, other than the fundamental units, but derived from these.

Design

NEMA design letters A, B and C define certain starting and running characteristics of three phase squirrel cage induction motors. These characteristics include locked-rotor torque, locked-rotor current, pull-up torque, breakdown torque, slip at rated load, and the ability to withstand full-voltage starting.

Design Current (of a circuit)

The magnitude of the current (rms value for a.c.) to be carried by the circuit in normal service.

Design Failure Mode and Effects Analysis (DFMEA)

A method for evaluating a design for robustness against potential failures.

Design Load

The actual, expected load or loads that a device or structure will support in service.

Design Test

Tests done to equipment to verify the design meets certain established charactistics or standards.

Designated Agent

Any entity that performs actions or functions on behalf of the Transmission Provider, an eligible Customer or the Transmission Customer required under the Tariff.

Detector

An instrument to detect the unbalance in a bridge circuit.

Detent

A stop or other device (such as a pin or lever) on a ratchet wheel. Switch action is typified by a gradual increase in force to a position at which there is an immediate and marked reduction in force.

Deterministic Jitter
Reproducible jitter within a given system, under controlled conditions. Also known as bounded jitter.

Device
The items installed in boxes that help control and distribute current, such as switches, receptacles, timers, thermostats, and dimmers.

Device Control Point
Local keypad on device level to control the switchgear often combined with local or remote switch.

Devitrification
The process in which a glass (noncrystalline or vitreous solid) transforms to a crystalline solid

Diallyl Phthalate (DAP)
A thermosetting plastic that offers outstanding dimensional stability and resistance to most chemicals and chemical compounds. It is used in the production of connector housings.

Diamagnetism
A weak form of induced or nonpermanent magnetism for which the magnetic susceptibility is negative.

Dielectric
Non-conductor of electricity. An insulator. Substance in which an electric field gives rise to no net flow of electric charge but only to a displacement of charge.

Dielectric Constant
Relative permittivity. It is the ratio of the capacitance of a capacitor with the given material as dielectric, to the capacitance of the same capacitor with vacuum (or air) as the dielectric.

Dielectric Grease
A silicone based chemical compound used to seal and lubricate connections between medium voltage connectors such as cable termination elbows.

Dielectric Heating
A form of heating in which electrically insulating material is heated by being subjected to an alternating electric field. Results from energy being lost by the field to electrons within the at-

oms and molecules of the material.

Dielectric Loss

Loss occurring in the leakage resistance of the dielectric.

Dielectric Strength

The maximum voltage that a dielectric material can withstand, under specified conditions, without rupturing. It is usually expressed as volts/unit thickness. Also called Disruptive Gradient or Electric Strength.

Dielectric Test

A test that is used to verify an insulation system. A voltage is applied of a specific magnatude for a specific period of time.

Dielectric Withstand

The ability of insulating materials and spacings to withstand specified overvoltages for a specified time (one minute unless otherwise stated) without flashover or puncture.

Dielectric Withstanding Voltage

Maximum potential gradient that a dielectric material can withstand without failure.

Difference Amplifier

A device that amplifies the difference between two inputs. It rejects any signals common to the two input.

Differential

The difference between operate AT and release AT. This is also expressed in % as (OAT-RAT)/OAT %.

Differential Nonlinearity (DNL)

A specification that appears in data-converter data sheets. In an ideal D/A converter, incrementing the digital code by 1 changes the output voltage by an amount that does not vary across the device's permitted range. Similarly, in an A/D, the digital value ramps smoothly as the input is linearly swept across its entire range. DNL measures the deviation from the ideal. An ideal converter has the code exactly the same size, and a DNL of 0 (zero).

Differential Remote Output Sensing

Uses a Kelvin connection at a remote location to sense the output voltage and better control the voltage at that point.

Differential Signaling

Most electrical signals are single-ended, comprised of a single wire and ground. Differential signals use two wires which are the inverse of each other — when one swings positive, the other swings negative in equal magnitude. The receiving circuit looks only at the difference between the two, ignoring any common-mode voltage. This "push-pull" arrangement reduces the impact of electrical interference because external noise will affect both wires equally and the common-mode rejection will ignore the noise.

Differentiator

An op amp whose output is proportional to the rate of change of the input signal.

Diffuse Insolation

Sunlight received indirectly as a result of scattering due to clouds, fog, haze, dust, or other obstructions in the atmosphere. Opposite of direct insolation.

Diffuse Radiation

The sunlight received indirectly, as a result of scattering due to clouds, fog, dust, moisture vapor or other substances in the atmosphere.

Diffuse Reflection

Diffusion by reflection in which, on the macroscopic scale, there is no regular reflection.

Diffuse Transmission

Transmission in which, on the macroscopic scale, there is no regular transmission.

Diffused Lighting

Lighting in which the light on the working plane or on an object is not incident predominantly from a particular direction.

Diffusion

Mass transport by atomic motion.

Diffusion Coefficient

The constant of proportionality between diffusion flux and the concentration gradient in Fick's first law.

Diffusion Furnace

Furnace used to make junctions in semiconductors by diffusing dopant atoms into the surface of the material.

Digital Meter

Show a discrete reading, in

the form of a decimal number, for a given input quantity.

Digital Multimeter (DMM)
Measuring instrument or VOM (e.g. voltage, resistance, current) with a digital display.

Digital Potentiometer (Digital Pot)
A solid-state device that emulates a mechanical potentiometer, it is usually controlled via a simple interface.

Digital Signal Processor
A microprocessor optimized in hardware design and software instruction set for the processing of analog signals digitally. This is achieved by DFT and similar techniques.

Dimmer
An electronic device with either a round knob, slide lever or finger-tip controlled buttons used to dim/brighten incandescent lighting. Available in a variety of wattages; fluorescent version also available.

Diode
The diode is the simplest and most fundamental nonlinear circuit element. It is a two terminal device which only allows current to flow in one direction.

Dip Tolerance (Lighting)
With regard to outdoor lighting, the percentage of instantaneous voltage variation from normal that is required to extinguish a light source.

Dipole (electric)
A pair of equal yet opposite electrical charges that are separated by a small distance

Direct Access
The ability of a retail customer to purchase commodity electricity directly from the wholesale market rather than through a local distribution utility.

Direct Beam Radiation
Radiation received by direct solar rays. Measured by a pyrheliometer with a solar aperture of 5.7° to transcribe the solar disc.

Direct Current (DC)
A type of electricity transmission and distribution by which electricity flows in one

direction through the conductor, usually relatively low voltage and high current. To be used for typical 120 volt or 220 volt household appliances, DC must be converted to alternating current, its opposite.

Direct Current (DC) Potentiometer

A potentiometer in which the supply is a battery and the balance is under d.c. conditions.

Direct Energy Conversion

Production of electricity from an energy source without transferring the energy to a working fluid or steam. For example, photovoltaic cells transform light directly into electricity. Direct conversion systems have no moving parts and usually produce direct current.

Direct Gain

In direct-gain buildings, sunlight directly enters the home through the windows and is absorbed and stored in massive floors or walls. These buildings are elongated in the east-west direction, and most of their windows are on the south side. The area devoted to south windows varies throughout the country. It could be as much as 20% of the floor area in sunny cold climates, where advanced glazings or moveable insulation are recommended to prevent heat loss at night. These buildings have high insulation levels and added thermal mass for heat storage.

Direct Insolation

Sunlight falling directly upon a collector. Opposite of diffuse insolation.

Direct Lighting

Lighting by means of luminaires with a light distribution such that 90 to 100 per cent of the emitted luminous flux reaches the working plane direct, assuming that this plane is unbounded

Direct Utility Cost

A utility cost that is identified with one of the DSM program categories (i.e., Energy Efficiency, Direct Load Control, Interruptible Load, Other Load Management, Other DSM Programs, Load Building).

Directional Relay

A protection relay in which the tripping decision is dependent in part upon the direction in which the measured quantity is flowing.

Direct-on-Line

A method of motor starting, which full line voltage is applied to a stationary motor.

Discharge

Electrical discharge can occur by the release of the electric charge stored in a capacitor through an external circuit. It can also occur by the breakdown of gaseous dielectrics within solid dielectrics on the application of a field.

Discharge Factor

A number equivalent to the time in hours during which a battery is discharged at constant current usually expressed as a percentage of the total battery capacity, i.e., C/5 indicates a discharge factor of 5 hours. Related to discharge rate.

Discharge Lamp

Lamp in which the light is produced, directly or indirectly, by an electric discharge through a gas, a metal vapor, or a mixture of several gases and vapors.

Discharge Rate

The rate, usually expressed in amperes or time, at which electrical current is taken from the battery.

Disconnect

Switch gear used to connect or disconnect components in a photovoltaic system.

Disconnect Switch

A simple switch that is used to disconnect an eletrical circuit. It may or may not have the ability to stop the flow of current in the circuit.

Disconnecting Means

A device or group of devices, or other means whereby all the ungrounded conductors of a circuit can be disconnected simultaneously from their source of supply.

Disconnector

A mechanical switching device which, in the open position, complies with the requirements specified for isolation. A disconnector is otherwise known as an isolator.

Discrimination

The ability of a power protection system to differentiate between the conditions it was intended to operate and those it was not intended for.

Dislocation

A linear crystalline defect around which there is an atomic misalignment.

Dispatchability

This is the ability of a generating unit to increase or decrease generation, or to be brought on line or shut down at the request or a utility's system operator.

Displacement Current

The current flows in a circuit containing a capacitor whenever the capacitor charges or discharges.

Dissipation

Loss of electric energy as heat.

Distributed Energy Resources (DER)

A variety of small, modular power-generating technologies that can be combined with energy management and storage systems and used to improve the operation of the electricity delivery system, whether or not those technologies are connected to an electricity grid.

Distributed Generation

A distributed generation system involves small amounts of generation located on a utility's distribution system for the purpose of meeting local peak loads and/or displacing the need to build additional local distribution lines.

Distributed Power

Generic term for any power supply located near the point where the power is used. Opposite of central power.

Distribution

The process of delivering electricity through low-voltage power lines to a consumer's home or business. Distribution includes local wires, transformers, substations and other equipment used to deliver electricity from the high-voltage transmission lines to homes and businesses.

Distribution Automation

A system consisting of line equipment, communications infrastructure, and information technology that is used to gather intelligence about a distribution system. It provides analysis and control in order to optomize operating efficency and reliability. The system can include distribution, small substation, and transmission line feeder reclosers, regulators and sectionalizers, which can be remotely monitored and controlled.

Distribution Board

An assembly containing switching or protective devices (e.g. fuses, circuit breakers, residual current operated devices) associated with one or more outgoing circuits fed from one or more incoming circuits, together with terminals for the neutral and protective circuit conductors. It may also include signalling and other control devices. Means of isolation may be included in the board or may be provided separately.

Distribution Charge

As electric utility services are unbundled, utilities will separate the fees associated with each major component of electricity service, including generation, transmission, and distribution. For MPSC-regulated electric utilities, the MPSC approves the rates that utilities charge customers for distribution services. A utility distribution charge is a charge for delivering electricity from a customer's chosen supplier to their residence or business, and may include a customer charge, demand charge, or energy charge.

Distribution Equipment

A device designed to provide electricity to multiple connections.

Distribution Line

This is a line or system for distributing power from a transmission system to a consumer. It is any line operating at less than 69,000 volt.

Distribution System

A term used to describe that part of an electric power system that distributes the electricity to consumers from a bulk power location such as a substation. It includes all lines and equipment beyound the substation fence.

Distribution Transformer

A transformer that reduces

voltage from the supply lines to a lower voltage needed for direct connection to operate consumer devices.

Distribution Voltage

A nominal operating voltage of 1-38kV.

Distributive Power

A packaged power unit located at the point of demand. While the technology is still evolving, examples include fuel cells and photovoltaic applications.

Diversion Load

Water and wind turbines require diversion loads to use the excess power they generate after the battery bank is fully charged. Ventilation fans and heating elements are popular choices.

Diversity Exchange

An exchange of capacity or energy, or both, between systems whose peak loads occur at different times.

Diversity Factor

The ratio of the sum of the maximum power demands of the subdivisions, or parts of a system, to the maximum demand of the whole system or of part of the system under consideration.

Divestiture

The stripping off of one utility function from the others by selling (spinning-off) or in most other way changing the ownership of the assets related to that function. Most commonly associated with spinning-off generation assets so they are no longer owned by the shareholders that own the transmission and distribution assets.

Door

The fuse tube of a fused cutout.

Dopant

A chemical element (impurity) added in small amounts to an otherwise pure semiconductor material to modify the electrical properties of the material. An n-dopant introduces more electrons. A p-dopant creates electron vacancies (holes).

Dot Notation

A notation used to denote similar ends of mutually coupled coils.

Double Arming Bolt

A special long bolt used to assemble two crossarms, one on each side of the pole.

Down Converters

A device which provides frequency conversion to a lower frequency, e.g. in digital broadcast satellite applications.

Downtime

Time when the photovoltaic system cannot provide power for the load. Usually expressed in hours per year or that percentage.

Downwind

Refers to a Horizontal Axis Wind Turbine in which the hub and blades point away from the wind direction, the opposite of an Upwind turbine.

Drag

In a wind generator, the force exerted on an object by moving air. Also refers to a type of wind generator or anemometer design that uses cups instead of a blades with airfoils.

Drain

Withdrawal of current from a cell.

Draught Tube

The flared tube bridging the gap between a reaction-style water turbine and the tail water. A draught tube maintains the sealed system necessary for creating 'suction head' which dramatically increases turbine power output.

Drawing

A deformation technique used to fabricate metal wire and tubing. Deformation is accomplished by pulling the material through a die by means of a tensile force applied on the exit side.

Draw-Lead

A cable or solid conductor that has one end connected to the transformer or a reactor winding and the other end drawn through the bushing hollow tube and connected to the top terminal of the bushing.

Drop

The voltage drop developed across a resistor due to current flowing through it.

Drop-Out

A relay drops out when it moves from the energized position to the un-energized position.

Dry Cell

A primary cell in which the electrolyte is absorbed in a porous medium, or is otherwise restrained from flowing. Common practice limits the term "dry cell" to the Leclanch, cell, which is the common commercial type.

Dry Charge (Battery)

The process by which the electrodes are formed and assembled in a charged state. The cell or battery is activated when electrolyte is added.

Dry Location

A location not normally subject to dampness or wetness. A location classified as dry may be temporarily subject to dampness or wetness, as in the case of a building under construction.

Dry-Type Tranformers

Transformers that use only dry-type materials for insulation. These have no oils or cooling fluids and rely on the circulation of air about the coils to provide necessary cooling. Such units are usually limited in size to a few hundred kVA because of problems of cooling the larger units.

Dual Channel (Dual-trace) Oscilloscope

An oscilloscope that has two independent input connectors and vertical sections and can display them simultaneously.

Dual Phase Controller

Switching regulator that employs dual-phase technique to reduce output noise and boost output current capability.

Dual Voltage Switch

A switch used to select primary windings of a transformer.

Dual Voltage Transformer

A transformer that has switched windings allowing its use on two different primary voltages.

Dual-Band

Dual-band refers to the capability of GSM network infrastructure and handsets to operate across two frequency bands.

Duality Principle

The duality principle establishes an analogy between similar variables when governed by analogous differential equations. For example, series circuits and parallel circuits are analogous if resistance is replaced by conductance, conductance by resistance, inductance by capacitance or vice versa, current by voltage or vice versa. There is also an analogy between electrical and magnetic circuits.

Dual-Modulus Prescaler

A Dual-Modulus Prescaler (DMP) is an important circuit block used in frequency synthesizers to divide the high-frequency signal from the voltage controlled oscillator (VCO) to a low-frequency signal by a predetermined divide ratio, either (N+1) or N, which is controlled by a swallow counter.

This low-frequency signal is then further divided by the main counter to the desired channel-spacing frequency which is then fed to the phase detector to form the closed feedback loop in frequency synthesizers.

Duct

A channel for holding and protecting conductors and cables, made of metal or an insulating material, usually circular in cross section like a pipe. Also referred to as Conduit.

Duct Bank

Two or more ducts or conduits used as part of a system.

Ductility

A measure of a material's ability to undergo appreciable plastic deformation before fracture.

Duplex

An adapter that provides two female receptacle openings when plugged into a single receptacle opening.

Dustproof

Constructed or protected so that dust will not interfere with its successful operation.

Dusttight

Constructed so that dust will not enter the enclosing case under specified test conditions.

Duty

A continuous or short-time rating of a machine. Continuous-duty machines reach an equilibrium temperature within the temperature limits of the insulation system. Machines which do not, or cannot, reach an equilibrium temperature have a short-time or intermittent-duty rating. Short-time ratings are usually one hour or less for motors.

Duty Cycle

The ratio of the lengths of the positive to negative halves of a waveform expressed in percentages.

Duty Rating

The amount of time an inverter (power conditioning unit) can produce at full rated power.

Dwelling

One or more rooms for the use of one or more persons as a housekeeping unit with space for eating, living, and sleeping, and permanent provisions for cooking and sanitation.

Dynamic Contact Resistance (DCR)

The electrical resistance of closed contacts, when the contact is in continuous operation under load.

Dynamic Pressure

The water pressure in a pipeline while the water is flowing. It is equal to the static pressure (measured in a closed pipline) minus pressure loss from friction, turbulence and cavitation in the pipeline and fittings.

Dynamic Range

The range, in dB, between the noise floor of a device and its defined maximum output level.

Dynamo

Device for converting mechanical energy into electrical energy. The mechanical energy of rotation is converted into electrical energy in the form of a current in the armature.

Dynamometer Instrument

This instrument is also a moving coil instrument except that in this case, the permanent magnet is replaced by a pair of fixed coils to give the fixed field.

Dyne

Unit of force in the c.g.s. system. 1 dyne = 10^{-5} N

Earth

The conductive mass of the Earth, whose electric potential at any point is conventionally taken as zero.

Earth Electrode

A conductor or group of conductors in intimate contact with, and providing an electrical connection to, Earth.

Earth Electrode Resistance

The resistance of an earth electrode to Earth.

Earth Fault Current

A fault current which flows to Earth.

Earth Fault Loop Impedance

The impedance of the earth fault current loop starting and ending at the point of earth fault. The earth fault loop comprises the following, starting at the point of fault: the circuit protective conductor, and the consumer's earthing terminal and earthing conductor, and for TN systems, the metallic return path, and for TT and IT systems, the earth return path, and the path through the earthed neutral point of the transformer, and the transformer winding, and the phase conductor from the transformer to the point of fault.

Earth Fault Protection

A protection system which is designed to excite during faults to earth.

Earth Ground

A low impedance path to earth for the purpose of discharging lightning, static, and radiated energy, and to maintain the main service entrance at earth potential.

Earth Leakage

Flow of current from a live

conductor to earth in an unintended path through the insulation.

Earth Leakage Circuit Breaker

The ELCB is designed to protect both equipment and users from fault currents between the live and earth conductors by detecting the rise in voltage of the frame earth connection with respect to a reference earth.

Earthed Concentric Wiring

A wiring system in which one or more insulated conductors are completely surrounded throughout their length by a conductor, for example a metallic sheath, which acts as a PEN conductor.

Earthed Equipotential Zone

A zone within which exposed conductive parts and extraneous conductive parts are maintained at substantially the same potential by bonding, such that, under fault conditions, the differences in potential between simultaneously accessible exposed and extraneous conductive parts will not cause electric shock.

Earthing

Connection of the exposed conductive p" of an installation to the main earthing terminal of that installation. Earthing conductor. A protective conductor connecting the main earthing terminal of an installation to an earth electrode or to other means of earthing.

Earthing Transformer

A three-phase transformer intended essentially to provide a neutral point to a power system for the purpose of grounding.

East Central Area Reliability Coordination Agreeme

One of the ten regional reliability councils that make up the North American Electric Reliability Council (NERC).

Echo

Effect produced when sound is reflected from a surface sufficiently far away for the reflected sound to be separately distinguishable.

Econo Reset

The simplest form of microprocessor supervisory circuit, it monitors the power supply for the microprocessor and provides only a power-on reset function.

Econ Oscillator

Low-cost, surface-mount, CMOS oscillator family from Dallas Semiconductor. EconOscillators replace crystal-based oscillators. They need no external crystals or timing components.

##Eddy Current

Circulating current produced in connecting materials by a varying magnetic field. Eddy currents are undesirable in the core of a transformer.

Eddy Current Loss

Power loss in magnetic materials due to eddy currents. This loss is proportional to the square of the thickness and hence can be reduced by the use of laminations.

Edge-Defined Film-Fed Growth (EFG)

A method for making sheets of polycrystalline silicon in which molten silicon is drawn upward by capillary action through a mold.

Edison Base

An internally-threaded lampholder, with the inner shell approx. 1" in diameter. Designed for widely-used standard medium base lamps.

Effective Earthing

Effective earthing avoids having dangerous potentials on the equipment even during electrical faults and also ensures the proper operation of electrical protection equipment during fault conditions. A system is said to be effectively earthed if the factor of earthing does not exceed 80%, and non-effectively earthed if it does.

Effective Internal Resistance (Battery)

The apparent opposition to current within a battery that manafests itself as a drop in battery voltage proportional to discharge current. Its value is dependent on battery design, state-of-charge, temperature and age.

Effective Number of Bits (ENOB)

An indication of the quality of an analog-to-digital converter (ADC). The measurement is related to the test frequency and the signal-to-noise ratio.

Effective Range

The range of values of the characteristic quantity or quantities. For example the energizing quantities to which the relay will respond and satisfy the requirements to precision.

Effective Series Resistance (ESR)

Effective Series Resistance (or Equivalent Series Resistance or ESR) is the resistive component of a capacitor's equivalent circuit.

A capacitor can be modeled as an ideal capacitor in series with a resistor and an inductor. The resistor's value is the ESR.

Effective Value

The value of an alternating current that produces the same heating effect in a pure resistance as a corresponding value of dc. The effective value of a sine curve is equal to .707 times its peak value. Also called "Root Means Squared (RMS) Value".

Effective/Equivalent Series Inductance (ESL)

Effective/Equivalent Series Inductance is the parasitic inductance in a capacitor or resistor.

Effectively Grounded

Intentionally connected conductors or electric equipment to earth, where the connection and conductors are of sufficently low impedance to allow the conducting of an intended current.

Efficacy

Relative ability to produce a desired effect. The amount of energy service delivered per unit of energy input.

Efficiency

The percentage of input power available for used by the load. The mathematical formula is: **Efficiency = P_o / P_i** Where "P_o" equals power output, "P_i" equals power input, and power is represented by watts.

Electrical Degrees

One cycle of AC. power is divided into 360 degrees. This allows mathematical relationships between the various aspects of electricity. Also, what the mothers of many liberal arts majors wish their daughters had married (or vice-versa)

Efficiency (Lighting)

A ratio of light emitted from

a luminaire to the light produced by the bare lamp.

Efficiency of a Machine

The ratio of the output energy to the input energy, usually expressed as a percentage. The efficiency of a machine can never exceed unity or 100%.

Efficiency Service Company

A company that offers to reduce a client's electricity consumption with the cost savings being split with the client.

Elastic Modulus or Modulus of Elasticity

The ratio of the stress to the strain in a given material.

Elasticity of Demand

The ratio of the percentage change in the quantity demanded of a good to the percentage change in price.

Elastomer

A material that at room temperature stretches under low stress to at least twice its length and snaps back to original length upon release of stress.

Elbow

A device used to connect a medium voltage cable (4-35KV nominal) to an electrical component such as a switch or transformer. Its name is derrived from the fact that its shape is an "L". Elbows are available in ratings of 200, 600 and 900 Ampere and are manufactured by the Elastimold division of Thomas & Betts.

Electric Capacity

This refers to the ability of a power plant to produce a given output of electric energy at an instant in time, measured in kilowatts or megawatts (1,000 kilowatts).

Electric Charge

Charge is an electrical property of the atomic particles of which matter is made. The elementary particle called the electron is negatively charged while the proton is equally positively charged so that normal matter is electrically neutral. [Unit: coulomb or C]

Electric Circuit

Path followed by electrons from a power source (generator or battery) through an external line (including devices that use the electricity) and returning through another line to the source.

Electric Company
In a deregulated market, the company that delivers electricity to a customer's home or business through its system of poles, power lines and other equipment.

Electric Current
An electric current flows through a conductor when there is an overall movement of charge through it and is measured as the time rate of change of charge.

Electric Distribution Company
The company that owns the power lines and equipment necessary to deliver purchased electricity to the consumer.

Electric Energy
The flow of charged particles (electrons).

Electric Field
The space near a charged BODY where other charges are affected. Similar to the gravitational field near a planet, except that it can also repel. The term is also used to describe how the field will affect other charges (which way and how much it will accelerate them).

Electric Flux
A measure of the electricity coming out from a charged surface. [Unit: coulomb]

Electric Flux Density
Electric flux passing through unit area perpendicular to the direction of the flux. [Unit: C/m^2]

Electric Light
Illumination produced by the use of electricity. The light sources may be of the incandescent, fluorescent, gas discharge or LED type.

Electric Meter
Generally, a device that measures the amount of electricity a customer uses. The primary types of electric meters are energy meters, demand meters, interval demand meters, and time-of-use meters. An energy meter is the simplest type of electric meter. It measures electricity use, referred to as kilowatt-hours. A demand meter measures kilowatt-hours used, and also the maximum electric use referred to as peak capacity or load. An interval

demand meter records the demand used in each measuring period. The periods are typically every 15 minutes, half-hour, or hour, depending on the specific meter and the way that the utility rates are calculated. Some utilities offer "time-of-use" rates, where customers pay different charges for electricity used during different times. The price might vary depending on the time of day, week, season, or year, or even depending on the hourly market price of electricity. A time-of-use meter measures customer electricity use and sometimes demand and records that data, along with the time of day, so the utility can bill the customer according to the charges established in the customer's time-differentiated rates.

Electric Motor
Device for converting electrical energy into mechanical energy in the form of rotation.

Electric Plant (Physical)
A facility containing prime movers, electric generators, and auxiliary equipment for converting mechanical, chemical, and/or fission energy into electric energy.

Electric Polarization
A type of polarisation occurring in a dielectric.

Electric Power
Electric power is given by the product of the potential difference and the current.

Electric Power Supplier
Non-utility provider of electricity to a competitive marketplace.

Electric Rate Schedule
A statement of the electric rate and the terms and conditions governing its application, including attendant contract terms and condi-

tions that have been accepted by a regulatory body with appropriate oversight authority.

Electric Reliability Council of Texas (ERCOT)

One of the ten regional reliability councils that make up the North American Electric Reliability Council (NERC).

Electric Resistance Heating

A type of heating system that generates heat by passing current through a conductor, causing it to heat up. These systems usually use baseboard heaters, often with individual controls. They are inefficient and are best used as a backup to more efficient options, such as solar heating or a heat pump.

Electric Service Provider

An entity that provides electric service to a retail or end-use customer.

Electric Shock

A dangerous physiological effect resulting from the passing of an electric current through a human body or livestock. Injury to the skin or internal organs that results from exposure to an electrical current. Electric shock occurs when the body becomes a part of an electric circuit. The electrical current must enter the body at one point and leave at another. The human body is a good conductor of electricity. Direct contact with electrical current can be potentially fatal. While some electrical shocks may appear not to be serious, there still may be serious internal damage, especially to the heart and brain.

Electric Strength

The maximum potential gradient that a material can withstand without rupture. Also called Dielectric Strength and Disruptive Gradient.

Electric System

This term refers to all of the elements needed to distribute electrical power. It includes overhead and underground lines, poles, transformers, and other equipment.

Electric Vehicle (EV)

EVs are generally cars, trucks or vans that performs the same functions as a conventional fueled vehicle—highway speeds, safety, etc. However, batteries are the storage means for the "fuel" and the vehicle is refueled directly from electricity by plugging into a special charger or wall socket. An EV can be created by converting a vehicle (such as a conventional gas car, to be powered by an electric motor and controllers) or it may have been produced directly from a company as an EV.

Electric Welding

In electrical welding, a very high electric current produces the heat needed to melt the material and join two metals together.

Electric Wholesale Generator

A power producer who sells power at cost to a customer.

Electrical Degree

One cycle in a rotating electric machine is accomplished when the rotating field moves from one pole to the next pole of the same polarity. There are 360 electrical degrees in this time period. [i.e. for each pair of poles there are 360 electrical degrees. In a machine with more than one pair of

poles, one electrical cycle is completed for each pair of poles in the mechanical cycle; or the electrical degrees per revolution is obtained by multiplying the number of pairs of poles by 360.]

Electrical Equipment

Any item for such purposes as generation, conversion, transmission, distribution or utilisation of electrical energy, such as machines, transformers, apparatus_ measuring instruments, protective devices, wiring systems, accessories, appliances and luminaires.

Electrical Fault

An abnormal connection causing current to flow from one conductor to ground or to another conductor. A fault may be corrected automatically or may lead to a voltage sag or power outage.

Electrical Grid

An integrated system of electricity distribution, usually covering a large area.

Electrical Hazard

A dangerous condition such that contact or equipment

failure can result in electric shock, arc flash burn, thermal burn, or blast.

Electrical Heating

The heating characteristic of an electric current is used extensively in industrial and domestic heating applications. Electric heating can be obtained from (a) resistance heating, (b) induction heating, (c) eddy current heating (d) dielectric heating, and (e) electric arc heating.

Electrical Horsepower

746 watts.

Electrical Installation

An assembly of associated electrical equipment supplied from a common origin to fulfil a specific purpose and having certain coordinated characteristics.

Electrical Noise

An unwanted electrical signal that produces an undesirable effect.

Electrical Panel

A box or panel where circuit breakers or fuses are located.

Electrical Relay

A device designed to produce sudden predetermined changes in one or more electrical circuits after the appearance of certain conditions in the controlling circuit.

Electrical Safety

Recognizing hazards associated with the use of electrical energy and taking precautions so that hazards do not cause injury or death.

Electrical Units

In the practical system, electrical units comprise the volt, the ampere, the ohm, the watt, the watt-hour, the coulomb, the Henry, the mho, the joule, and the farad.

Electrically Independent Earth Electrodes

Earth electrodes located at such a distance from one another that the maximum current likely to flow through one of them does not significantly affect the potential of the other(s).

Electrician/Electrical Contractor

A person who specializes in the knowledge of electricity and the prescribed electrical codes.

Electricity

A form of energy produced by the flow of particles of matter and consists of commonly attractive positively (protons [+]) and negatively (electrons [-]) charged atomic particles. A stream of electrons, or an electric current.

Electricity (or Power) Marketer

An entity that obtains energy from any source or combination of sources (including independent generators, utility system power or spot purchases) for delivery to a utility or consumer.

Electricity Supplier

An entity that has been licensed by the Public Service Commission to sell electricity to customers within the State of Maryland.

Electro Magnetic Interference (EMI)

Any electromagnetic disturbance, phenomenon, signal or emission that causes or can cause an undesired response, malfunction, or degradation of performance of electrical and electronic equipment.

Electro-Absorption Modulators (EAM)

Chip-level modulation devices often integrated into hybrid transponder devices, alongside lasers.

Electrochemical Breakdown

In a practical insulation ions may arise from dissociation of impurities or from slight ionisations of the insulating material itself. When these ions reach the electrodes, reactions occur in accordance with Faraday's law of electrolysis, but on a much smaller scale. The products of the electrode reaction may be chemically or electrically harmful and in some cases can lead to rapid failure of the insulation.

Electrochemical Cell

A device containing two conducting electrodes, one positive and the other negative, made of dissimilar materials (usually metals) that are immersed in a chemical solution (electrolyte) that transmits positive ions from the negative to the positive electrode and thus forms an electrical charge. One or more cells constitute a battery.

Electrochemical Couple

The system of active materials within a cell that provides electrical energy storage through an electrochemical reaction.

Electrocution

The destruction of life by means of electric current.

Electrode

An electrical conductor through which an electric current enters or leaves a conducting medium, whether it be an electrolytic solution, solid, molten mass, gas, or vacuum. For electrolytic solutions, many solids, and molten masses, an electrode is an electrical conductor at the surface of which a change occurs from conduction by electrons to conduction by ions. For gases and vacuum, the electrodes merely serve to conduct electricity to and from the medium.

Electrodynamometer

An instrument dependant on the interaction of the electromagnetic fields of fixed and movable coils. It can measure current, voltage or power in both d.c. and a.c.

Electrolier

Similar to the Edison Medium Base lampholder, but with a smaller outer diameter.

Electrolysis

Electric current passing through an electrolyte which produces chemical changes in it.

Electrolyte

A liquid conductor of electricity. In batteries, usualy H2SO4, sulfuric acid, but may be any number of things. Seawater is the most common electrolyte in the world and by suspending a zinc and a steel sheet in it, you can get a little electricity.

Electrolyte (Battery)

In a lead-acid battery, the electrolyte is sulfuric acis diluted with water. It is a conductor and also a supplier of hydrogen and sulfate ions for the reaction.

Electrolytic Capacitor (Condenser)

An electrical capacitor in which one electrode is a metal foil coated with a thin layer of the metal oxide, and the other electrode is a non-corrosive salt paste. The metal foil is maintained positive to prevent the removal of the oxide film by the hydrogen liberated.

Electromagnet

A magnet produced by passing an electric current through and insulated wire conductor coiled around a core of soft iron, as in the fields of a dynamo or motor.

Electromagnetic Compatibility (EMC)

The ability of electronic equipment to be a "good electromagnetic neighbor": It neither causes, nor is susceptible to, electromagnetic interference (within the limits of applicable standards).

Electromagnetic Damping

Electromagnetic damping is produced by the induced effects when the coil moves in the magnetic field and a closed path is provided for the currents to flow.

Electromagnetic Field

Electric and magnetic force field that surrounds a moving electric charge.

Electromagnetic Induction

The process of developing a voltage in a wire that is being either cut by or is cutting a magnetic field.

Electromagnetic Interference (EMI)

A term that describes electrically induced noise or transients, usually at frequencies above 1 MHz.

Electromagnetic Relay

A relay controlled by electromagnetic means, to open and close electric contacts.

Electromagnetic Spectrum

The range of frequencies over which electromagnetic radiations are propagated.

Electromechanical

A mechanical device which is controlled by an electric device. Solenoids and shunt trip circuit breakers are examples of electromechanical devices.

Electromotive Force (Emf)

The source of electrical energy required to produce an electric current in a circuit. Defined as the rate at which electrical energy is drawn from the source and dissipated in a circuit when unit current is flowing in the circuit. [Unit: volt or V]

Electron

An elementary particle of an atom with a negative electrical charge and a mass of 1/1837 of a proton; electrons surround the positively charged nucleus of an atom and determine the chemical properties of an atom. The movement of electrons in an electrical conductor constitutes an electric current.

Electron Affinity

The tendency of an atom or molecule to accept an electron and form a negative ion. The halogens have high electron affinities.

Electron Emission

The escape of electrons from certain materials.

Electron Flow

Electrical current is the flow of electrons. It is produced when an electron from one atom knocks electrons of another atom out of orbit. When an atom contains only one valence electron, that electron is easily given up when struck by another electron. The striking electron gives its energy to the electron being struck. The striking electron settles into orbit around the atom, and the electron that was struck moves off to strike another electron. This same effect in the game of pool. If the moving cue ball strikes a stationary ball. The stationary ball then moves off with the most of the cue ball's energy, and the cue ball stops moving. The stationary ball did not move off with all the energy of the cue ball. It moved off with most of the energy of the cue ball. Some of the cue ball's energy was lost to heat when it struck the stationary ball. Some energy is also lost when one electron strikes another. That is why a wire heats when current flows through it. If too much current flows through a wire, overheating will damage the wire and possibly become a fire hazard.

Electron Gun

The source of electrons in a cathode ray tube. Consists of a cathode emitter of electrons, an anode with an aperture through which the beam of electrons can pass, and one or more focussing and control electrodes.

Electron Lens

A system of electric or magnetic fields used to focus a beam of electrons in a manner analogous to an optical lens.

Electron Tube

An arrangement of two or more conductive elements, enclosed in an envelope, to control electron flow in a circuit.

Electron Volt (eV)

The amount of kinetic energy gained by an electron when accelerated through an electric potential difference of 1 Volt; equivalent to 1.603×10^{-19}; a unit of energy or work.

Electronegativity

For an atom, having a tendency to accept valence electrons.

Electronic Ballasts

An electronic device that regulates the voltage of fluorescent lamps. Compared to older magnetic ballasts, electronic ballasts use less electricity and are not prone to the flickering and humming effects sometimes associated with magnetic ballasts.

Electronic Equipment

Equipment constructed with electronic chips. Common examples include TVs, radios, stereos and computers.

Electronic Industries Alliance (EIA)

Among other things, the EIA sponsors electrical and electronic standards.

Electropositivity

The degree to which an element in a galvanic cell will function as the positive element of the cell. An element with a large electropositivity will oxidize faster than an element with a smaller electropositivity.

Electrostatic

A Potential difference (electric charge) measurable between two points which is caused by the distribution if dissimilar static charge along the points. The voltage level is usually in kilovolts (volts times 1000).

Electrostatic Discharge

Release of stored static electricity. Most commonly: The potentially damaging discharge of many thousands of volts that occurs when an electronic device is touched by a charged body.

Electrostatic Generator

A machine designed for the continuous separation of electric charge. An example is the Van de Graaf Generator.

Electrostatic Meter

These basically work on the principle that the force (or torque) of attraction is proportional to the product of the charges and the force is proportional to the square of the voltage. Thus this meter reads the mean square value and hence is calibrated to read the root mean square value. The electrostatic meter is basically a voltmeter.

Electrostatic Precipitator

An electronic pollution-control device that removes particles of fly ash from a power plant's waste gases.

Elevation

The sun's angle above the horizon.

Ellipsoid

A solid figure traced out by an ellipse rotating about one of its axes.

Elongation

The amount (% length) that a conductor or other material can stretch before breaking when a pulling force is applied.

Embedded Cost

A utility's average cost of doing business, which includes the costs of fuel, personnel, plants, poles, and wires.

Embedded Generation

(dispersed generation or distributed generation) Plant which is connected directly to (embedded within) the utility's distribution network rather than to the high voltage transmission system (or nation grid). They are generally considered to be less than 10-100 MW in capacity and are not centrally planned or dispatched. They are commonly found on industrial sites and in areas of high renewable energy source such as wind, hydro and solar.

Embedded System

A system in which the computer (generally a microcontroller or microprocessor) is included as an integral part of the system. Often, the computer is relatively invisible to the user, without obvious applications, files, or operating systems. Examples of products with invisible embedded systems are the controller that runs a microwave oven or the engine control system of a modern automobile.

Emergency Lighting
Lighting provided for use when the supply to the normal lighting fails.

Emergency power
An independent reserve source of electric power which, upon failure or outage of the normal power source, provides stand-by electric power.

Emergency Stopping
Emergency switching intended to stop an operation.

Emergency Switching
An operation intended to remove, as quickly as possible, danger, which may have occurred unexpectedly.

Eminent domain
The authority to acquire land from a private owner for the benefit of public use.

Emitter
The part of the transistor that is the source of carriers. For npn transistors, the emitter sends free electrons into the base, whereas for pnp transistors, the emitter sends holes into the base.

Empirical
Based upon the results of experiment and observation only.

Enclosure
The cabinet or specially designed box or fence or walls in which electrical controls and apparatus are housed, to prevent personnel from accidentally contacting energized parts or to protect the equipment from physical damage. A part providing protection of equipment against certain external influences and in any direction protection against direct contact.

Encoder
A digital circuit that converts information into coded form.

End Point
Behavior of the device at the limit of temperature or voltage.

End-of-Discharge Voltage

The voltage of a battery at the termination of a discharge but before the discharge is stopped.

End Point Voltage

The Cell or Battery voltage at which point the rated discharge capacity has been delivered at a specific Rate-of-Discharge. It is also used to specify the cell or battery voltage below which the connected equipment will not operate or below which operation is not recommended.

End-Use

The specific purpose for which electric is consumed (i.e. heating, cooling, cooking, etc.).

Energized Equipment

Equipment that has voltage potential available within it, such as an electrical panel when the main circuit breaker is switched on.

Energizing

Electrically connected to a source of potential difference.

Energy

The capacity for doing work as measured by the capability of doing work (potential energy) or the conversion of this capability to motion (kinetic energy). Energy has several forms, some of which are easily convertible and can be changed to another form useful for work. Most of the world's convertible energy comes from fossil fuels that are burned to produce heat that is then used as a transfer medium to mechanical or other means in order to accomplish tasks. Electrical energy is usually measured in kilowatthours, while heat energy is usually measured in British thermal units.

Energy Charge or Electric Charge

The charge for the electricity used by an electric customer during the billing period, measured in kilowatt-hours.

Energy Contribution Potential

Recombination occurring in the emitter region of a photovoltaic cell.

Energy Costs

Costs, such as for fuel, that are related to and vary with energy production or consumption.

Energy Deliveries

Energy generated by one electric utility system and delivered to another system through one or more transmission lines.

Energy Density

Ratio of cell energy to weight or volume (watt-hours per pound, or watt-hours per cubic inch).

Energy Efficiency Ratio (EER)

The ratio of the cooling capacity of the air conditioner, in Btu per hour, to the total electrical input in watts under test conditions specified by the Air-Conditioning and Refrigeration Institute.

Energy Levels

The energy represented by an electron in the band model of a substance.

Energy Management System

A system in which a dispatcher can monitor and control the flow of electric power by opening and closing switches to route electricity or to isolate a part of the system for maintenance. It is also used to control the amount of generation needed to serve a load.

Energy Payback Time

The time required for any energy producing system or device to produce as much energy as was required in its manufacture. For solar electric panels, this is about 16-20 months.

Energy Receipts

Energy generated by one electric utility system and received by another system through one or more transmission lines.

Energy Reserves

The portion of total energy resources that is known and can be recovered with presently available technology at an affordable cost.

Energy Resources

Everything that could be used by society as a source of energy.

Energy Saving Devices

Devices utilized within a dwelling designed to more efficiently make use of energy sources while providing heating, cooling, and light.

Energy Services Companies (ESCOs)

ESCOs would be created in

a deregulated, openly competitive electric marketplace. The Energy Services industry would be made up of power aggregators, power marketers and brokers, whose job is to match buyers and sellers, tailor both physical and financial instruments to suit the needs of particular customers, and to allow even the smallest residential customers to form buying groups or cooperatives that will give them the same bargaining power as large industrial customers.

Energy Source
The primary source that provides the power that is converted to electricity through chemical, mechanical, or other means. Energy sources include coal, petroleum and petroleum products, gas, water, uranium, wind, sunlight, geothermal, and other sources.

Energy Use
Energy consumed during a specified time period for a specific purpose (usually expressed in kWh).

Engine
A device for converting one form of energy into another, especially for converting other forms of energy into mechanical (kinetic) energy.

Engineering Ceramics
Technical ceramics for structural applications.

Enhanced Data Rates for GSM Evolution (EDGE)
An enhanced modulation technique designed to increase network capacity and data rates in GSM networks. EDGE should provide data rates up to 384Kbps.

Ethylene Propylene Dione Monimer (EPDM)
This is a synthetic rubber compound used as insulation in making electrical components.

Entrance Cable/Service Entrance Conductor
This is the cable running down the side of a customer's house into the meter. This cable is owned by the customer and its maintenance is the customer's responsibility. Work on this cable should be performed only by a licensed electrician.

Environment
All the natural and living

things around us. The earth, air, weather, plants, and animals all make up our environment.

Environmental Attributes

Environmental attributes quantity the impact of various options on the environment. These attributes include particulate emissions, SO2 or Nox, and thermal discharge (air and water).

Epitaxial Growth

The growth of one crystal on the surface of another crystal. The growth of the deposited crystal is oriented by the lattice structure of the original crystal.

Equalization

The process of restoring all cells in a battery to an equal state-of-charge. Some battery types may require a complete discharge as a part of the equalization process.

Equalizing Charge

A controlled overcharge of an already full battery to restore all the individual cells within the battery to the same state of charge.

Equilateral

Having all the sides equal in length. Equilateral triangle is one which has all three sides equal.

Equilibrium

State of balance between opposing forces or effects.

Equinox

The two times of the year when the sun crosses the equator and night and day are of equal length; usually occurs on March 21st (spring equinox) and September 23 (fall equinox).

Equipment

A general term including materials, fittings, devices, appliances, fixtures, apparatus, and the like used as a part of, or in connection with, an electrical installation.

Equipotential Bonding

Electrical connection maintaining various exposed conductive parts and extraneous conductive parts at substantially the same potential.

Equipotential Lines and Surfaces

Lines and surfaces having the same electric potential.

Equity Capital

The sum of capital from retained earnings and the issuance of stocks.

erg

Unit of work or energy in the c.g.s. system of units. 1 erg = 10^{-7} J

Erosion

In a surface discharge, if the products of decomposition are volatile and there is no residual conducting carbon on the surface, the process is simply one of pitting and is known as erosion. Erosion occurs in organic materials.

Error

An error is a deviation from the true value of the measured variable.

Error Vector Magnitude (EVM)

A measure of the difference between the (ideal) waveform and the measured waveform. The difference is called the error vector, usually referred to with regard to M-ary I/Q modulation schemes like QPSK, and shown on an I/Q "constellation" plot of the demodulated symbols.

Escape Provision

A contract provision which allows a party, such as an electric customer, to get out of it. Usually, there is a penalty.

Ethylene Propylene Rubber (EPR)

A synthetic rubber compound that is used as cable insulation.

Ethylene Vinyl Acetate (EVA)

An encapsulant used between the glass cover and the solar cells in PV modules. It is durable, transparent,resistant to corrosion, and flame retardant.

Eutectic

An alloy used to form the melting point of a fuse. It is frequently silver or tin based.

Eutectic Phase

One of the two phases found in the eutectic structure.

Evaluation Kit

Evaluation Kit (EV Kit, Development Kit): A printed circuit board with a Maxim/ Dallas* IC and support components. Most Evaluation Kits are fully assembled and tested.

Evaluation System (EVSYS)

Evaluation kits that also include an interface board for connecting to a personal computer and Windows-based EVKit software.

EVSYS: Suffix used for Maxim Evaluation System part numbers.

exa (E)

Decimal multiple prefix corresponding to 10^{18}

exbi (Ei)

Binary multiple prefix corresponding to gigabinary or 2^{60} or $(2^{10})^6$ or 1024^6. [IEC 1998]

Excitation

The addition of energy to a nucleus, an atom or a molecule transferring it from its ground state to a higher energy level. The excitation is the difference in energy between the ground state and the excited state.

Exciting Current

The magnatizing current of a device such as a transformer. Also known a field current.

Explosion Proof

Designed and constructed to withstand and internal explosion without creating an external explosion or fire.

Exponent

The number indicating the power of a quantity.

Export of Electrical Energy

Supply of Electrical Energy by a Generator to the CEB system.

Exposed Conductive Part

A conductive part of equipment which can be touched and which is not a live part but which may become live under fault conditions.

Exposed Pad

Offered in some packages to improve thermal dissipation or lower the impedance of the ground connection. Normally not electrically isolated, it typically needs to be connected to a ground or power plane, depending on the device.

Expressway Roadway (Lighting)

A divided major roadway for through-traffic with partial control of access and generally with interchanges at major crossroads. Expressways for non-commercial traffic within parks and park-like area are generally known as parkways.

Extended Superframe (ESF)

A DS1 framing format in which 24 DS0 times lots, plus a coded framing bit are organized into a frame which is repeated 24 times to form a superframe.

Extension, Box

An add-on section that fits to the bottom or to the top of a gread level box, extending its height.

Extra High Voltage

An electrical system or cable designed to operate at 345kv (nominal) or higher.

Extraneous Conductive Part

A conductive part liable to introduce a potential, generally earth potential, and not forming part of the electrical installation.

Extrapolation

Filling in values or terms of a series on either side of the known values thus extending the range of values.

Extrinsic Semiconductor

A semiconductor in which the carrier density results mainly from the presence of impurities or other imperfections, as opposed to an intrinsic semiconductor in which the electrical properties are characteristics of the ideal crystal.

Extrusion

A forming technique whereby a material is forced, by compression, through a die orifice.

F

1. Farad(s): Unit of capacitance
2. f in lower case is the standard abbreviation for femto, a metric prefix for 10 to the -15.

Face-centered Cubic (FCC)

A crystal structure found in some of the common elemental metals. Within the cubic unit cell, atoms are located at all corner and face-centered positions.

Factor of Earthing

This is the ratio of the highest r.m.s. phase-to-earth power frequency voltage on a sound phase during an earth fault to the r.m.s. phase-to-phase power frequency voltage which would be obtained at the selected location without the fault. This ratio characterises, in general terms, the earthing conditions of a system as

viewed from the selected fault location.

Factory Acceptance Test (FAT)

Validation procedures witnessed by the customer at the factory.

Fahrenheit

Temperature scale in which the melting point of ice is taken as 32 °F and the boiling point of water under standard atmospheric pressure (760 torr) as 212 °F. A Fahrenheit degree is 1/180 of the difference between these two temperatures.

Fail-Safe

A technique used in RS-485 interface transceivers which forces the output to a predefined state in the event of a line short or open circuit.

Fan Controller-Linear

An integrated circuit that

varies the speed and airflow of a cooling fan using a variable voltage in response to temperature or system commands.

Fan Hanger

A single receptacle with a specialized cover plate that provides a hook or other means of supporting a wall fan.

Farad

The unit of measure for capacitance. It is the capacitance of a capacitor in which an applied voltage of one volt will store a charge of one coulomb. The more practical units of capacitance are the microfarad and picofarad.

Faraday

Quantity of electricity required to liberate or deposit 1 gram-equivalent of an ion. 1 Faraday = 96,490 coulomb.

Faraday Cage

The name given to a device that shields its inside from electric fields generated by static electricity. Usually a complete conductive shell, it collects stray charges and, because like charges repel, stores them on the outside surface (where they can be further apart than on the inside). The electric fields generated by these charges then cancel each other out on the inside of the cage. Often used to protect sensitive radio equipment.

Fast Fourier Transform (FFT)

An algorithm for converting data from the time domain to the frequency domain; often used in signal processing.

Fast Neutrons

Neutrons resulting from nuclear fission that have lost little of their energy by collision and therefore travel at high speeds. It is usual to define neutrons with energies in excess of 0.1 MeV as fast.

Fatigue

Failure, at relatively low stress levels, of structures that are subjected to fluctuating and cyclic stresses.

Fault

A circuit condition in which current flows through an abnormal or unintended path. This may result from ah insulation failure or a bridging of insulation. Con-

. . . .

ventionally the impedance between live conductors or between live conductors and exposed or extraneous conductive parts at the fault position is considered negligible.

Fault Blanking

A function that ignores a fault for a predetermined period. This is done to eliminate nuisance fault indication.

Fault Close Rating

The ability, in amps, of a switching device to "close" into a fault of specific magnatude, without excessive arcing.

Fault Current

The current that flows as a result of a short-circuit condition.

Fault Indicator

A device installed on a conductor to determine if current exceeded the indicator's current rating. Fault indicators sense using use the magnetic field induced by load current.

Fault Protection System

A system composed of devices that are able to sense abnormal current flow and prevent it. PGE's fault protection system includes circuit breakers, fuses, reclosers and switches.

Fault Tolerant

Will tolerate excessive voltage during a fault condition.

Feasibility Factor

A factor used to adjust potential energy savings to account for cases where it is impractical to install new equipment. For example, certain types of fluorescent lighting require room temperature conditions. They are not feasible for outdoor or unheated space applications. Some commercial applications, such as color-coded warehouses, require good color rendition, so color distortions could also make certain types of lighting infeasible. The feasibility factor equals 100 percent minus the percent of infeasible applications.

Federal Energy Reglatory Commission (FERC)

FERC is an independent regulatory agency within the U.S. Department of Energy that approves rates for wholesale electricity transactions and transmission of electricity in interstate commerce for utilities, power marketers, power pools, power exchanges and independent system operators. FERC also regulates the transmission and sale for resale of natural gas in interstate commerce; regulates the transmission of oil by pipeline in interstate commerce; licenses and inspects private, municipal and state hydroelectric projects; and oversees related environmental matters. The FERC board of governors is composed of five commissioners. The chairman, designated by the President, serves as the commission's administrative head. FERC is based in Washington, D.C.

Feedback

The term is generally applied to electronic amplifiers to which a portion of the output energy is used to reduce or increase the amplification, by reacting on an earlier stage according to the relative phase of the return.

Feeder

This is an electrical supply line, either overhead or underground, which runs from the substation, through various paths, ending with the transformers. It is a distribution circuit, usually less than 69,000 volts, which carries power from the substation.

Feeder Lockout

This happens when a main circuit is interrupted at the substation by automatic protective devices and cannot be restored until crews investigate. This indicates a serious problem on the circuit, usually equipment failure or a broken conductor.

Feed-Through

An in-line switch that can be attached at any point on a length of flexible cord to provide switching control of attached equipment.

Femto (f)

Decimal sub-multiple prefix corresponding to 10^{-15}.

Femtoampere (fA)

Femtoampere(s): 10 to the -15 Ampere; a millionth a nanoampere.

Fermi

A unit of length used in

nuclear physics. 1 fermi = 10^{-15} meter.

Fermi Energy
For a metal, the energy corresponding to the ·highest filled electron state in the valence bond at 0 K.

Fermi Level
Energy level at which the probability of finding an electron is one-half. In a metal, the Fermi level is very near the top of the filled levels in the partially filled valance band. In a semiconductor, the, Fermi level is in the band gap.

Fermi-Dirac Statistics
The branch of statistical mechanics used with systems of identical particles which have the property that their wave function changes sign if any two particles are interchanged.

Ferrimagnetism
The type of magnetism occurring in materials in which the magnetic moments of adjacent atoms are anti-parallel, but of unequal strength, or in which the number of magnetic moments oriented in one direction outnumber those in the reverse direction. Typical ferrimagnetic materials are the ferrites.

Ferrites
A group of ceramic materials which exhibit the property of ferrimagnetism. As they are basically electrical insulators, they do not suffer from the effects of eddy currents.

Ferroelectrics
Dielectric materials which exhibit properties such as hysteresis which are usually properties of ferromagnetic materials.

Ferromagnetism
Permanent and large magnetizations found in some metals (e.g., fe, ni, and co), which result from the parallel alignments of neighboring magnetic moments.

Ferroresonance
In transformers, an over-voltage condition that can occur when the core is excited through capacitance in series with the inductor. This is especially prevalent in transformers that have very low core losses. It can generally be prevented by having a load connected to

the transformer secondary. Contact Young & Company or Howard Industries for additional information.

Ferroresonant Transformer
A voltage regulating transformer which depends on core saturation and output capacitance.

Fertile Material
Isotopes which can be transformed into fissile material by the absorption of neutrons.

Fiber
Any material that has been drawn into a cylinder with a length-to-diameter ratio greater than about ten.

Fiber Optics
Piping light is the science that deals with the transmission of light through extremely thin fibers of glass, plastic, or other transparent material.

Fiber-to-the-home (FTTH)
A method for broadband data (voice, Internet, multimedia, etc.) delivery to the home via optical fiber. Contrast with FTTN (fiber-to-the-node) which uses fiber up to a node outside the home and uses copper to bring the data into the home.
There are two technologies for delivering broadband: Fiber-to-the-node (FTTN) uses fiber to bring data to a node and uses copper to bring the data into the home. Fiber-to-the-home (FTTH) brings fiber all the way into the home.

Fibre Channel
A highly-reliable, gigabit interconnect technology that allows concurrent communications among workstations, mainframes, servers, data storage systems, and other peripherals using SCSI and IP protocols. It provides interconnect systems for multiple topologies that can scale to a total system bandwidth on the order of a terabit per second. (The standardized spelling is "fibre channel" but often misspelled as "fiber channel.")

Fiducial Value
A specified value to which reference is made in order to specify the accuracy of the transducer. For transducers the fiducial value is the span. For transducers having reversible or sym-

metrical outputs the fiducial value can be either the span or half the span as specified by the manufacturer.

Field

The windings of an electric generator which are supplied with dc to produce the steady electromagnetic field. Generators used for demonstration purposes may use permanent magnets to produce the magnetic field.

Field Coil

A coil of wire used for magnetising an electromagnet.

Field Current

The magnatizing current of a device such as a transformer. Also known a exciting current.

Field Emission

The emission of electrons from an unheated surface as a result of a strong electric field existing at that surface.

Field Programmable Gate Array (FPGA)

A family of general-purpose logic devices that can be configured by the end user to perform many, different, complex logic functions. It is often used for prototyping logic hardware.

Field-Effect Transistor (FET)

A transistor in which the voltage on one terminal (the gate) creates a field that allows or disallows conduction between the other two terminals (the source and drain). There are two varieties: The JFET (Junction Field-Effect Transistor) and the MOSFET (Metal-Oxide-Semiconductor Field-Effect Transistor). The FET is one of two major kinds of transistor, the other being the Bipolar Junction Transistor.

Filament

A thin thread. A wire of high melting point heated by the passage of current inside a vacuum tube, incandescent lamp or other similar device.

Fill

In conduit or cable tray installations, the portion of the total cross-sectional area of the tray or conduit that can be occupied by conductors or cables

Fill Factor

The ratio of a photovoltaic

cell's actual power to its power if both current and voltage were at their maxima. A key characteristic in evaluating cell performance.

Filler
A material used in multiconductorcable to occupy large interstices formed by the cable assembly. Also, a material added to an insulation compound to add volume and increase impact resistance

Filter
A circuit that is designed to pass signals with desired frequencies and reject or attenuate other frequencies.

Filter Frequency Range
The frequency range within which the filter operates.

Final Circuit
The final circuits in an electrical wiring system. A circuit connected directly to current using equipment, or to a socket outlet or socket outlets or other outlet points for the connection of such equipment.

Financial Attributes
Financial attributes measure the financial health of the company. Utility management, security analysts, investors, and regulators use these attributes to evaluate a utility's performance against its historic records and industry averages. Key financial attributes include capital requirements, earnings per share of common equity, capitalization ratios, and interest coverage ratios.

Firing
A high-temperature heat treatment that increases the density and strength of a ceramic piece.

Firm Energy
Power or power-producing capacity covered by a commitment to be available at all times during the period.

Firm Gas
Gas sold on a continuous and generally long-term contract.

Firm Power
Power or power producing capacity intended to be available at all times during the period covered by a guaranteed commitment to deliver, even under adverse conditions.

Firm Transmission Service

Point-to-point transmission service that is reserved and/or scheduled for a term of one year or more and that is of the same priority as that of the Transmission Provider's firm use of the transmission system. Firm Transmission service that is reserved and/or scheduled for a term of less than one year shall be considered Short-Term Firm Transmission Service for the purposes of service liability.

First-In First Out (FIFO)

A type of memory that stores data serially, where the first bit read is the first bit that was stored.

Fish Tape

A coiled spring-steel line used for pulling, or fishing, cable and wire through enclosed spaces.

Fissile Material

Isotopes which are capable of undergoing nuclear fission. Sometimes the term is restricted to apply only to isotopes which are capable of undergoing fission upon impact with a slow neutron.

Fission Products

Both stable and unstable isotopes produced as a result of nuclear fission.

Fitting

An accessory such as a locknut, bushing, or other part of a wiring system that is intended primarily to perform a mechanical rather than an electrical function.

Fixed Capacitor Bank

A capacitor bank installed with no automatic switching device. The bank is manually switched on and off.

Fixed Equipment

Equipment designed to be fastened to a support or otherwise secured in a specific location.

Fixed Tilt Array

A photovoltaic array set in at a fixed angle with respect to horizontal.

Fixture

With regard to lighting, a reference to Luminaire.

Flame

Glowing mass of gas produced during combustion.

Flame Resistance
The ability of insulation or jacketing material to resist the support and conveyance of fire.

Flanged Inlet
A plug intended for flush mounting on appliances or equipment to provide a means for power connection via a cord connector.

Flanged Outlet
A receptacle intended for flush mounting on appliances or equipment to provide a means for power connection via an inserted plug.

Flash ADCs
An analog-to-digital converter that uses a series of comparators with different threshold voltages to convert an analog signal to a digital output.

Flash Hazard
A dangerous condition associated with the release of energy caused by an electric arc.

Flash Hazard Analysis
A study investigating a worker's potential exposure to arc-flash energy, conducted for the purpose of injury prevention, the determination of safe work practices, and the appropriate levels of PPE.

Flash Point
The lowest temperature at which a substance gives off sufficient inflammable vapour to produce a momentary flash when a small flame is applied.

Flash Protection Boundary
An approach limit at a distance from exposed live parts within which a person could receive a second degree burn if an electrical arc flash were to occur.

Flash Suit
A complete FR clothing and equipment system that covers the entire body, except for the hands and feet. This includes pants, jacket, and bee-keeper-type hood fitted with a face shield.

Flashover
Flashing due to high current flowing between two points of different potential. Usually due to insulation breakdown resulting from arcing.

Flat Rate
A fixed charge for goods

and services that does not vary with changes in the amount used, volume consumed, or units purchased.

Flat-Plate Array

A photovoltaic (PV) array that consists of non-concentrating PV modules.

Flat-Plate Module

An arrangement of photovoltaic cells or material mounted on a rigid flat surface with the cells exposed freely to incoming sunlight.

Flat-Plate Photovoltaic Module

An arrangement of photovoltaic cells mounted on a rigid flat surface with the cells exposed freely to incoming sunlight.

Flat-Plate PV

Refers to a PV array or module that consists of non-concentrating elements. Flat-plate arrays and modules use direct and diffuse sunlight, but if the array is fixed in position, some portion of the direct sunlight is lost because of oblique sun-angles in relation to the array.

Fleming's Rules

If the forefinger, second fin-ger, and thumb of the right hand are extended at right angles to each other, the forefinger indicates the direction of flux, the second finger the direction of the emf and the thumb the direction of the motion in an electric generator. If the left hand is used, the fingers indicate the conditions for an electric motor.

Flexible Cable

A cable whose structure and materials make it suitable to be flexed while in service.

Flexible Cord

A flexible cable in which the cross sectional area of each conductor does not exceed 4 mm^2.

Flexible Fuel Vehicle

A vehicle that can run on a variety of fuels (most often just gasoline and E85)

Flexible Load Shape

The ability to modify your utility's load shape on short notice. When resources are insufficient to meet load requirements, load shifting or peak clipping may be appropriate.

Flexible Retail Pool

A model for the restructured

electric industry that features an Independent System Operator (ISO) operating in parallel with a commercial Power Exchange, which allows end-use consumers to buy from a spot market or "pool" or to contract directly with a particular supplier.

Flexible Wiring System

A wiring system designed to provide mechanical flexibility in use without degradation of the electrical components.

Float Charge

A method of maintaining a cell or battery in a charged condition by continuous, long-term, constant voltage charging at a level sufficient to balance self-discharge.

Float Charging

Method of recharging in which a secondary cell is continuously connected to a constant-voltage supply that maintains the cell in fully charged condition. Typically applied to lead acid batteries.

Float Life

Number of years that a battery can keep its stated capacity when it is kept at float charge.

Float Service

A battery operation in which the battery is normally connected to an external current source; for instance, a battery charger which supplies the battery load< under normal conditions, while also providing enough energy input to the battery to make up for its internal quiescent losses, thus keeping the battery always up to full power and ready for service.

Float-Zone Process

In reference to solar photovoltaic cell manufacture, a method of growing a large-size, high-quality crystal whereby coils heat a polycrystalline ingot placed atop a single-crystal seed. As the coils are slowly raised the molten interface beneath the coils becomes a single crystal.

Flooded Cell

A cell design that incorporates an excess amount of electrolyte.

Fluctuation

A surge or sag in voltage

amplitude, often caused by load switching or fault clearing.

Flue Gas Desulfurization Unit (Scrubber)

Equipment used to remove sulfur oxides from the combustion gases of a boiler plant before discharge to the atmosphere. Chemicals, such as lime, are used as the scrubbing media.

Flue Gas Particulate Collectors

Equipment used to remove fly ash from the combustion gases of a boiler plant before discharge to the atmosphere. Particulate collectors include electrostatic precipitators, mechanical collectors (cyclones), fabric filters (baghouses), and wet scrubbers.

Flume

Open and closed flumes serve to channel water into a reaction-type water turbine.

Fluorescence

That property by virtue of which certain solids and fluids become luminous under the influence of radiant energy.

Fluorescent Lamps

Fluorescent lamps produce light by passing electricity through a gas, causing it to glow. The gas produces ultraviolet light; a phosphor coating on the inside of the lamp absorbs the ultraviolet light and produces visible light. Fluorescent lamps produce much less heat than incandescent lamps and are more energy efficient. Linear fluorescent lamps are used in long narrow fixtures designed for such lamps. Compact fluorescent light bulbs have been designed to replace incandescent light bulbs in table lamps, floodlights, and other fixtures.

Fluorescent Starter

A device with a voltage-sensitive switch and a capacitor that provides a high-voltage pulse to start a fluorescent lamp. Rated in watts.

Flush

A wallplate designed for flush-mounting with wall surfaces or the plane surfaces of electrical equipment.

Flush-Mounted

An inlet intended to be installed flush with the surface

of a panel or a piece of equipment.

Flux
The lines of force of a magnetic field.

Flux Linkage
The linking of the magnetic lines of force with the conductors of a coil. The value obtained by multiplying the number of turns in the coil by the number of magnetic lines of force passing through the coil.

Fluxmeter
An instrument for the measurement of magnetic flux.

Fly Ash
Particle matter from coal ash in which the particle diameter is less than 1×10^{-4} meter. This is removed from the flue gas using flue gas particulate collectors such as fabric filters and electrostatic precipitators.

Focus
The oscilloscope control that converges the CRT electron beams to produce a sharp display.

Foldback Current Limit
A circuit which reduces the current limit once the device enters current-limited operation. Commonly seen on RS-422/RS-485 drivers and some power circuits.

Foot (ft)
Imperial unit of length. 1 foot = 12 inches = 304.8 mm exactly

Foot Candle
Unit of illumination at a point one foot distance from a one candela source. (in the imperial system of units) 1 foot candle = 1 lumen per square foot.

Foot Lambert
Unit of luminance. It is the luminance of a uniform diffuser emitting a foot candle.

Foot Pound Second (FPS) System
The foot pound second system of units is an imperial set of units derived from the fundamental units of the foot, the pound mass and

the second.

Force

An elementary physical cause capable of modifying the motion of a mass.

Forced Outage

The shutdown of a generating unit, transmission line or other facility, for emergency reasons or a condition in which the generating equipment is unavailable for load due to unanticipated breakdown.

Forced-Air (FA)

Forced-Air, a cooling classification for transformers now classified as ONAF. Oil type, Forced circulation through cooling (i.e. cooling pumps) and natural convection flow in windings.

Force-Sense

Measurement technique in which a voltage (or current) is forced at a remote point in a circuit; then the resulting current (or voltage) is measured (sensed).

Forebay

A closed tank at the top end of a hydro power diversion pipeline. It allows the water to settle before entering the penstock. Usually where the primary filter/trash-rack is installed.

Forging

Mechanical forming of a metal or alloy by heating and hammering.

Form Factor

Ratio of the rms value to the average value in a periodic waveform.

Fortin Barometer

A mercury in glass barometer, which used in conjunction with correction tables enables accurate measurement of atmospheric pressure to be made.

Forward Bias

A dc voltage applied to a PN junction semiconductor so that the positive terminal of the voltage source connects to the P-type material and the negative terminal to the N-type material. It produces forward current in the circuit.

Forward Converter

A power-supply switching circuit that transfers energy to the transformer secondary when the switching transistor is on.

Forward Current

Current in a circuit of a semiconductor device due to conduction by majority carriers across the PN junction.

Forward Error Correction (FEC)

A technique for detecting and correcting errors from imperfect transmission by adding a small number of extra bits. FEC allows optical transmission over longer distances by correcting errors that can happen as the signal-to-noise ratio decreases with distance.

Forward Transfer Impedance

The amount of impedance placed between the source and load with installation of a power conditioner. With no power conditioner, the full utility power is delivered to the load; even a transformer adds some opposition to the transfer of power. On transformer based power conditioners, a high forward transfer impedance limits the amount of inrush current available to the load.

Fossil Fuel

Materials such as coal, oil or natural gas used to produce heat or power; also called conventional fuels. These materials were formed in the ground millions of years ago from plant and animal remains.

Fossil Fuel Plant

A plant using coal, oil, gas and other fossil fuel as its source of energy.

Fossil Fuels

Fuels formed in the ground from the remains of dead plants and animals. It takes millions of years to form fossil fuels. Oil, natural gas, and coal are fossil fuels.

Fossil-Fuel Plant

A plant using coal, petroleum, or gas as its source of energy.

Fourier Analysis

The expansion of a mathematical function or an experimentally obtained waveform in the form of a trigonometric series.

Fourier Series

Resolution of a periodic function into its direct component, its fundamental sinusoidal component and an infinite series of harmonic sinusoidal components.

Fourier Transform

An integral transformation from the time domain to the frequency domain.

Four-In-One or "Quad"

A receptacle in a common housing that accepts up to four plugs. Four-In-One receptacles can be installed in place of duplex receptacles mounted in a single-gang box, providing a convenient means of adding receptacles without rewiring.

Four-Way

A switch used in conjunction with two 3-Way switches to control a single load (such as a light fixture) from three or more locations. This switch has four terminal screws and no ON/OFF marking.

Four-way Switch

A set of three switches wired to control the same fixture or group of fixtures.

Framer

A device used to align/synchronize to an embedded framing pattern in a serial bit stream. Once synchronized and data fields are properly aligned, overhead bits for alarms, performance monitoring, embedded signaling, etc. may be extracted and processed.

Francis Turbine

A type of reaction turbine. Francis turbines have nine or more fixed vanes on the runner. Water enters the runner from the side (through the vertical vanes), and exits out the bottom of the turbine (a 90 degree change in direction). Francis turbines operate with 4 to 2000 feet of head, and can be as large as 800 megawatts of output.

Free Electron
An electron which is not attached to an atom, molecule or ion, but is free to move under the influence of an electric field.

Free Energy
A thermodynamic quantity that is a function of both the internal energy and entropy of a system.

Free-running Trace
A trace that is displayed without being triggered and either with or without an applied signal.

Freezing Point
The temperature of equilibrium between solid and liquid substance at a pressure of one standard atmosphere.

Frenkel Defect
In an ionic solid, a cation-vacancy and cation-interstitial pair.

Frequency
In an AC system, the value of voltage and current rise from zero to a maximum, falls to zero, increases to a maximum in the opposite direction, and falls back to zero again. This complete set of values is called a cycle. The number of complete cycles passed through in one second is called the frequency. The General Conference on Weights and Measures has adopted the name hertz (abbreviated Hz) as the unit of frequency. The common power frequency in North America is 60 Hz. In Europe and most of Africa and Asia it is 50 Hz. Airplanes typically use 400 Hz systems.

Frequency (Noise) Attenuation
The range of attenuation (limiting) for a given frequency range. In this case, the greater the negative number, the more noise reduction.

Frequency Bin
A band of frequencies of a specific width.

Frequency Counter
A circuit that can measure

and display the frequency of a signal.

Frequency Hopping Spread Spectrum (FHSS)

A transmission technology in which the data signal is modulated by a narrowband carrier signal which changes frequency ("hops") over a wide band of frequencies. The hopping seems random but is prescribed by an algorithm known to the receiving system.

Frequency Modulation (FM)

A process whereby the frequency of the carrier is controlled by the modulating signal.

Frequency Regulation

This indicates the variability in the output frequency. Some loads will switch off or not operate properly if frequency variations exceed 1%.

Frequency Response

The frequency response of a circuit is the variation of its behaviour (voltage or current) with change in frequency.

Frequency Shift Keying

A method of transmitting

digital data by shifting the frequency of a carrier signal to represent binary 1s and 0s.

Frequency Spectrum

The frequency spectrum of a signal consists of the plots of the amplitude and phases of the harmonics against frequency.

Frequency Synthesizer

A frequency synthesizer is an electronic circuit that uses an oscillator to generate a preprogrammed set of stable frequencies with minimal phase noise. Primary applications include wireless/RF devices such as radios, set top boxes, and GPS.

Frequency Transducer

A transducer used for the measurement of the frequency of an A.C. electrical quantity.

Fresnel Lens

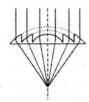

An optical device that focuses light like a magnifying glass; concentric rings are faced at slightly different angles so that light falling on any ring is focused to the same point. Fresnel lenses are flat rather than thick in the center and can be stamped out in a mold.

Fuel Adjustment

A clause in the rate schedule that provides for adjustment of the amount of a bill as the cost of fuel varies from a specified base amount per unit. The specified base amount is determined when rates are approved. This item is shown on all customer bills and indicates the current rate for any adjustment in the cost of fuel used by the company. It can be a credit or a debit. The fuel adjustment lags two months behind the actual price of the fuel. For example, the cost of oil in January will be reflected in March's fuel adjustment.

Fuel Adjustment Charge

A clause in the rate schedule that provides for adjustment of the amount of a bill as the cost of fuel varies from a specified base amount per unit. The speci-

fied base amount is determined when rates are approved. This item is shown on all consumer bills and indicates the current rate for any adjustment in the cost of fuel used by the utility. It can be a credit or a debit. The fuel adjustment lags usually lags one or two months behind the actual price of the fuel. For example, the cost of oil in January will be reflected in March's fuel adjustment.

Fuel Cell Vehicle (FCV)

Fuel Cell Vehicles (FCVs) have an electric drive train, but use a fuel cell, not the grid, to power the vehicle. Some of these vehicles also store a portion of their energy in batteries. Only prototype FCVs are currently available (under lease to governments or businesses). Hydrogen is stored in specialized tanks onboard the vehicle and vehicle refueling would take place at specialized refueling centers (that are not widely available today). Hydrogen is not technically a fuel, it is an energy carrier, and is the most abundant and simplest known element. The hydrogen to power a fuel cell must first be separated from other materials—water, biomass, natural gas, etc. It takes about four times more energy to separate the hydrogen from a source and then use it to power a vehicle compared to the energy used to power a BEV.

Fuel Cells
Devices that convert the chemical energy of fuels directly into electricity.

Fuel Diversity
A utility or power supplier that has power stations using several different types of fuel. Avoiding over-reliance on one fuel helps avoid the risk of supply interruption and price spikes.

Fuel Element
An element of nuclear fuel for use in a nuclear reactor, usually uranium encased in a case.

Fuel Escalation
The annual rate of increase of the cost of fuel, including inflation and real escalation, resulting from resource depletion, increased demand, etc.

Fuel Expenses
These costs include the fuel used in the production of steam or driving another prime mover for the generation of electricity. Other associated expenses include unloading the shipped fuel and all handling of the fuel up to the point where it enters the first bunker, hopper, bucket, tank, or holder in the boiler-house structure.

Fuel Mix
The proportions of each fuel type (e.g. nuclear, coal, oil, hydro, etc.) used by a power plant to generate electricity.

Fuel-Use Attributes
Fuel-use attributes are important to utilities concerned about reliance on a single fuel or reduction in usage of a particular fuel. These attributes include annual fuel consumption by type and percent energy generation by fuel.

Fulgurite
A glass-like structure that forms around the element of a current limiting fuse when it operates. It is causes when the heat of the arc melts the silica sand surrounding it.

Full Duplex
A channel providing simultaneous transmission in

both directions.

Full Duplex Communications
A communications system in which data can travel simultaneously in both directions.

Full Load Current
The largest current that a motor or other device is designed to carry under specific conditions. Also current at rated conditions.

Full Scale
The specified maximum magnitude of the input quantity being measured that can be applied to a transducer without causing a change in performance beyond specified tolerance

Full Scale Output
The specified maximum output value for which the stated accuracy condition applies

Full Sun
The amount of power density in sunlight received at the earth's surface at noon on a clear day (about 1,000 Watts/square meter).

Full Wave Rectifier
A rectifier with a centre

tapped secondary windings and two diodes, or a bridge rectifier circuit.

Full-Forced Outage
The net capability of main generating units that are unavailable for load for emergency reasons.

Full-Load Current
The current required for any electrical machine to produce its rated output or perform its rated function.

Full-Load Speed
The speed at which any rotating machine produces its rated output.

Full-Load Torque
The torque required to produce rated power at full-load speed.

Function Generator
A circuit that produces a variety of waveforms.

Functional Earth Conductor
Conductor to be connected to a functional earth terminal.

Functional Earth Terminal
Terminal directly connected to a point of a measuring supply or control circuit or

to a screening part which is intended to be earthed for functional purposes.

Functional Earthing

Connection to Earth necessary for proper functioning of electrical equipment.

Functional Equivalent (FE)

Functional equivalent (in component cross-reference data); also field engineer; also framing error

Functional Extra Low Voltage

An extra low voltage system in which not all of the protective measures required for SELV or PELV have been applied.

Functional Switching

An operation intended to switch 'on' or 'off' or vary the supply of electrical energy to all or part of an installation for normal operating purposes.

Functional Unbundling

The functional separation of generation, transmission, and distribution transactions within a vertically integrated utility without selling of "spinning off" these functions into separate companies.

Fundamental Displacement Factor (FDF)

Cosine of the phase difference between the fundamental components of voltage and current. For non distorted sinusoids, it is also equal to the power factor.

Fundamental Units

The units in which physical quantities are measured which are independent from each other.

Furling

The act of a wind generator Yawing out of the wind either horizontally or vertically to protect itself from high wind speeds.

Furling Tail

A wind generator protection mechanism where the rotor shaft axis is offset horizontally from the yaw axis, and the tail boom is both offset horizontally and hinged diagonally, thus allowing the tail to fold up and in during high winds. This causes the blades to turn out of the wind, protecting the machine.

Fuse

An over-current protective device for opening a circuit

by means of a conductor designed to melt and break when an excessive current flows along it for a sufficient time. The fuse comprises all the parts that form the complete device.

Fuse Arcing Time

The amount of time required to extinguish the arc and clear the circuit.

Fuse Carrier

The movable part of a fuse designed to carry a fuse link.

Fuse Element

A part of a fuse, which is designed to melt and thus open a circuit

Fuse Link

1. A replaceable fuse element used in a Fused Cutout. 2. A replacable part or assembly comprised entirely or principly of thr conducting element, requires to be replaced after each curcuit interruption to restore the fuse to operating conditions.

Fuse Melt Time

The time needed for a fuse element to melt, thereby initiating operation of the fuse. Also known as Melt Time.

Fused Cutout

A device, normally installed overhead, that is used to fuse a line or electrical apparatus.

Fusing Current

This is the minimum current that will cause the fuse element to heat up melt or blow.

Gain

A measure of amplification of a device, usually expressed in dB.

Gallium (Ga)

A chemical element, atomic number 31, metallic in nature, used in making certain kinds of solar cells and semiconductor devices.

Gallium arsenide (GaAs)

A semiconductor material used for optoelectronic products such as LEDs, and for high-speed electronic devices.

Gallon (Gal)

Imperial gallon. A measure of volume. 1 gal = 4.54596 litre

Galvanic Cell

A combination of electrodes, separated by electrolyte, that is capable of producing electrical energy by electrochemical action.

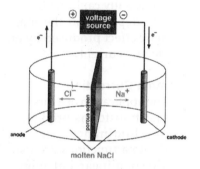

Galvanic Isolation

Refers to a design technique that separates signal current from AC power distribution introduced stray noise current.

Galvanised Iron

Iron coated with a layer of zinc to prevent corrosion, usually by hot dipping into the molten metal.

Galvanometer

An instrument for detecting, comparing, or measuring small electric currents, but

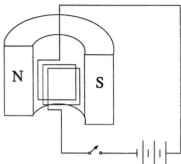

not usually calibrated. Usually depends on the magnetic effect produced by an electric current.

Gamma Correction

The application of a function that transforms brightness or luminance values. Gamma functions are usually nonlinear but monotonic and designed to affect the highlights (whitest values), midtones (grayscale), and shadows (dark areas) separately.

Most commonly applied to make a light-emitting device, such as a display, match the human eye's brightness curve. In other terms: A gamma correction function can be used to alter the luminance (light intensity) of a display such that its brightness (the human-perceived values) looks correct.

Gang

A term that describes the number of devices a wallplate is sized to fit (i.e. "2- gang" designates two devices).

Gas Turbine Plant

A plant in which the prime mover is a gas turbine. A gas turbine consists typically of an axial-flow air compressor, one or more combustion chambers, where liquid or gaseous fuel is burned and the hot gases are passed to the turbine and where the hot gases expand to drive the generator and are then used to run the compressor.

Gassing

The evolution of gas from one or both of the electrodes in a cell. Gassing commonly results from self-discharge or from the electrolysis of water in the electrolyte during charging.

Gassing (Battery)

The evolution of gas from one or more of the electrodes in a cell. Gassing commonly results from local action (self discharge) or from the electrolylis of water in the electrolyte during charging.

Gassing Current

Portion of charge current that goes into electrolytical production of hydrogen and oxygen from the electrolytic liquid. This current increases with increasing voltage and temperature.

Gate

The terminal of a FET that controls drain current. Also the terminal of a thyristor used to turn on the device.

Gateway

The Gateway is a computer which provides interfaces between the local computer system and one or several SCADA (or RCC) systems.

Gauge

The measure of the size of a wire. The smaller the number, the thicker the wire and the higher its current-carrying capacity.

Gauss (G)

An old unit for measuring magnetic flux density (or magnetic induction). 1 G = 10^{-4} T

Gaussian Minimum Shift Keying (GMSK)

Gaussian minimum shift keying (GMSK) is a form of frequency shift keying (FSK) used in GSM systems. The tone frequencies are separated by exactly half the bit rate. It has high spectral efficiency.

Gel

A chemical compound used to seal and mechanically cushion fiberoptic filament in a cable. The cleaners used to remove are made by American Polywater.

Gel Cleaner

A chemical based cleaner used to remove the gel in a fiberoptic cable. Gel cleaner is made by American Polywater Corp.

Gel-Type Battery

Lead-acid battery in which the electrolyte is composed of a silica gel matrix.

Generating Station (Generating Plant or Power Plan)

The location of prime movers, electric generators, and auxiliary equipment used for converting mechanical, chemical, and nuclear energy into electric energy.

Generating Unit

Any combination of physically connected generator(s), reactor(s), boiler(s), combustion turbine(s), or other prime mover(s) operated together to produce electric power.

Generation

The process of producing electrical energy. Generation may also refer to the amount of electrical energy produced, usually expressed in watt-hours, kilowatt-hours, or megawatthours (MWh).

Generation and Transmission Cooperative (G&T)

A power supply cooperative owned by a group of distribution cooperatives. G&Ts generate power or purchase it from public or investor-owned utilities, or from both.

Generation Charges

Part of the basic service charges on every customer's bill for producing electricity. Generation service is competitively priced and is not regulated by Public Utility Commissions. This charge depends on the terms of service between the customer and the supplier.

Generation Dispatch and Control

Aggregation and dispatch-ing (sending off to some location) generation from various generating facilities, providing backup and reliability services.

Generation Plant

A plant that has generators and other equipment for producing electricity.

Generator

A rotating machine which converts mechanical energy into electrical energy. In the automotive industry traditional terminology uses generator to refer to only those machines designed to produce dc current through brushes and a commutator (as opposed to alternator).

Generator Nameplate Capacity

The full-load continuous rating of a generator, prime mover, or other electric power production equipment under specific conditions as designated by the manufacturer. Installed generator nameplate rating is usually indicated on a nameplate physically attached to the generator.

Generator Step-Up (GSU)

Generator step up is done by transformers directly connected to the generator output terminals. This is usually done via busbars in large generating stations. They normally have a high voltage in secondary and high current in primary.

Geothermal

An electric generating station in which steam tapped from the earth drives a turbine-generator, generating electricity.

Geothermal Energy

Natural heat contained in the rocks, hot water and steam of Earth's subsurface; can be used to generate electricity and heat homes and businesses.

Geothermal Plant

A plant in which the prime mover is a steam turbine. The turbine is driven either by steam produced from hot water or by natural steam that derives its energy from heat found in rocks or fluids at various depths beneath the surface of the earth. The energy is extracted by drilling and/or pumping.

Gibi (Gi)

Binary multiple prefix corresponding to gigabinary or 2^{30} or $(2^{10})^3$ or 1024^3. [IEC 1998]

Giga (G)

Decimal multiple prefix corresponding to a billion(US) or 10^9.

Gigabit Interface Converter (GBIC)

A removable transceiver module permitting Fibre-Channel and Gigabit-Ethernet physical-layer transport.

Gigawatt

This is a unit of electric power equal to one billion watts, or one thousand megawatts—enough power to supply the needs of a medium-sized city.

Gilbert

The c.g.s. unit of magnetomotive force in electrostatic units. 1 gilbert = 10 A.

Glare

Condition of vision in which there is discomfort or a reduction in the ability to see significant objects, or both, due to an unsuitable distribution or range of luminance or to extreme contrasts in space or time.

Glass

An inorganic product of fusion which has cooled to a rigid condition without crystallizing.

Glass-ceramic

A fine-grained crystalline material that was formed as a glass and subsequently devitrified (crystallized).

Glazings

Clear materials (such as glass or plastic) that allow sunlight to pass into solar collectors and solar buildings, trapping heat inside.

Glitch

General term used to describe an undesirable, momentary pulse or unexpected input or output.

Glitch Immunity

A term used in microprocessor supervisory circuit datasheets to describe the maximum magnitude and duration of a negative-going V_{CC} supply-voltage pulse without causing the reset output to assert.

Global radiation

Total solar radiant energy impinging on a surface, equal to the sum of direct and diffuse radiation.

Global Warming

Global warming is the gradual increase in global temperatures caused by the emission of gases that trap the sun's heat in the Earth's atmosphere. Gases that contribute to global warming include carbon dioxide, methane, nitrous oxides, (CFC's) chlorofluorocarbons, and halocarbons (the replacements for CFC's). The carbon dioxide emissions are primarily caused by the use of fossil fuels for energy.

Glow Discharge

Electric discharge in which the secondary emission from the cathode is much greater than the thermionic emission.

Good Utility Practice

Any of the practices, methods, and acts engaged in or approved by a significant portion of the electric utility industry during the relevant time period, or any of the practices, methods, and acts which, in the exercise of reasonable judgment in light of the facts known at the time the decision was made, could have been expected to accomplish the desired result of the lowest reasonable cost consistent with good business practices, reliability, safety and expedition. Good Utility Practice is not intended to be limited to the optimum practice, method, or act to the exclusion of all others, but rather to be acceptable practices, methods, or acts generally accepted in the region and consistently adhered to by the Transmission Provider.

Goof Plate

An oversize cover plate designed to hide a rough patching job around a box. Note that drywall and plaster must be repaired to within 1/8" of any box; larger gaps may not simply be hidden behind a goof plate.

Governor

A device that regulates the speed of a rotating shaft, either electrically or mechanically.

Grain Boundaries

The boundaries where crystallites in a multicrystalline material meet.

Graph of Network

The geometric structure of the interconnection of the network elements which completely characterises the number of independent loop currents or the number of independent node-pair voltages necessary to study the network.

Graticule

The CRT grid lines that facilitate the location and measurement of oscilloscope traces.

Grease

1) Slang for Cable Pulling Lubricant, a chemical compound used to reduce pulling tension by lubricating a cable when pulled into a duct or conduit. 2) Slang for Dielectric Grease, a silicone based chemical compound used to seal and lubricate connections between medium voltage connectors such as cable termination elbows.

Green House Gas (GHG)

A gas that in the atmosphere prevents heat from radiating back into space, and thus warms the earth (the green-house effect). Carbon dioxide (CO_2) is the most common greenhouse gas. Methane (CH_4) is another GHG, and has approximately twenty times the greenhouse effect as CO_2 in the atmosphere.

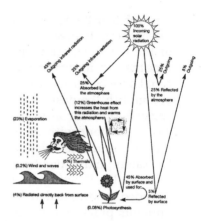

Green Power

A term used to describe electricity produced by sources that are less harmful to the environment than fossil fuels. While there is no strict definition of Green Power, generally renewable sources such as solar, wind power, geothermal, biomass, and small hydroelectric are considered Green Power sources.

Greenfield Plant

This refers to a new electric power generating facility built from the ground up.

Greenhouse Effect

The greenhouse effect allows solar radiation to penetrate but absorbs the infrared radiation returning to space. It thus increases the mean global surface temperature of the earth caused by gases in the atmosphere (including carbon dioxide, methane, nitrous oxide, ozone, and chlorofluorocarbon).

Greenhouse Gases

Gases that trap the heat of the sun in the Earth's atmosphere, producing the greenhouse effect; the two major greenhouse gases are water vapor and carbon dioxide; lesser greenhouse gases include methane, ozone, chlorofluorocarbons.

Grid

(power grid or utility grid) This refers to the public utility power system. If you get a monthly electric bill, you are "on the grid."

Grid Interconnection

A link between CEB Electricity system and Embedded Generator's Electricity System, made for the purpose of Exporting or Importing Electrical Energy.

Grid Lines

Metallic contacts fused to the surface of the solar cell to provide a low resistance path for electrons to flow out to the cell interconnect wires.

Grid System

An arrangement of power lines connecting power plants and consumers over a large area.

Grid-Connected (PV System)

A PV system in which the PV array acts like a central generating plant, supplying power to the grid.

Gross Generation

The total amount of electric energy produced by the generating units at a generating station or stations, measured at the generator terminals.

Ground

The ground is an arbitrarily decided point whose voltage is taken as zero. In many situations, equipment is connected physically to the actual, dirt ground, so that voltage is taken as zero—hence the name. In England the term "earth" is used, for the same reason. To be "grounded" means to be connected to a place that is maintained at the ground voltage.

Ground (Wire)

A conducting connection, whether intentional or accidental, between an electrical circuit or equipment and the earth, or to some conducting body that serves in place of the earth.

Ground Conductor

A conductive path used to connect equipment to a grounding electrode. A low impedance path for fault current to follow to help facilitate the operation of fault protection.

Ground Fault

An undesired current path between ground and an electrical potential.

Ground Fault Circuit Interrupters (GFCIs)

GFCIs can help prevent electrocution. They should be used in any area where water and electricity may come into contact. When a GFCI senses current leakage in an electrical circuit, it assumes a ground fault has occurred. It then interrupts power fast enough to help prevent serious injury from electrical shock. Test GFCIs according to the manufacturer's instructions monthly and after major electrical storms to make sure they are working properly. Replace all GFCIs that are not working properly, but never replace a GFCI with a standard non-GFCI outlet or circuit breaker. Do not use an appliance or device that trips a GFCI on a non-GFCI-protected circuit; instead, take the appliance to authorized repair center to be checked for faulty wiring or replace it.

Ground Fault Protection of Equipment

A system intended to provide protection of equipment from damaging line to ground fault currents by operating to cause a disconnecting means to open all ungrounded conductors of the faulted circuit. This protection is provided at current levels less than those required to protect conductors from damage through the operations of a supply circuit over current device.

Ground Loop

An undesirable feedback condition caused by two or more circuits sharing a common electrical line, usually a grounded conductor.

Ground State

The most stable energy state of a nucleus, atom or molecule.

Grounded

Connected to earth or to some conducting body that serves in place of the earth.

Grounded Conductor

A system or circuit conductor that is intentionally grounded, usually gray or white in color.

Grounded Neutral

The common neutral conductor of an electrical system which is intentionally connected to earth to provide a current carrying path for the line to neutral load devices.

Grounded, effectively

Intentionally connected to earth through a ground connection or connections of sufficiently low impedance and having sufficient current-carrying capacity to prevent the buildup of voltages that may result in undue hazards to connect equipment or to persons.

Grounding Conductor

A conductor used to connect metal equipment enclosures and/or the system grounded conductor to a grounding electrode, such as the ground wire run to the water pipe at a service; also may be a bare or insulated conductor used to ground motor frames, panel boxes, and other metal equipment enclosures used throughout electrical systems. In most conduit systems, the conduit is used as the ground conductor.

Grounding Electrode

The conductor used to connect the grounding electrode to the equipment grounding conductor, to the grounded conductor, or to both, of the circuit at the service equipment or at the source of a separately derived system.

Grounding Equipment Conductor

The conductor used to connect the non-current-carrying metal parts of equipment, raceways, and other enclosures to the system grounded conductor, the grounding electrode conductor, or both, of the circuit at the service equipment or at the source of a separately derived system.

Guide Vanes

Used in reaction turbines to change water flow direction by 90 degrees, causing the water to whirl and enter all turbine runner buckets simultaneously, improving turbine efficiency.

Guy Anchor

Attaches tower guy wires securely to the earth.

Guy Radius

The distance between a wind turbine tower and the guy anchors.

Guy Strain Insulator

An insulator, normally porcelain, used to electrically isolate one part of a down guy from another. Guy Strain Insulators are made by Porcelain Products.

Guy Wire

Attaches a tower to a Guy Anchor and the ground.

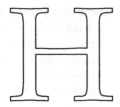

H

Henry(ries): The unit of inductance.

Half Wave Rectifier

A rectifier with only one diode in series with the load. The output is a half-wave rectified voltage with the other half wave being at zero voltage.

Half Wave Symmetry

A function has half-wave symmetry when one half of its waveform is exactly the negation of the previous half of the waveform.

Half-Duplex

Data transmission over a circuit capable of transmitting in either direction, but not simultaneously.

Half-Flash

An ADC architecture which uses a bank of comparators first to digitize the upper half bits, then uses a digital-to-analog converter (DAC) to subtract that voltage from the input, and then digitizes what remains of the input signal to get the lower half bits.

Half-Life

The time taken for the activity of a radioactive isotope to decay to half of its original value. In other words the time taken for half the atoms present to disintegrate.

Halide

Binary compound of one of the halogen elements (fluorine, chlorine, bromine or iodine).

Hall Effect

The phenomenon whereby a force is brought to bear on a moving electron or hole by a magnetic field that is applied perpendicular to the direction of motion. The force direction is perpendicular to both the magnetic field and the particle motion directions.

Halogen Lamp

Gas-filled lamp containing a tungsten filament and a small proportion of halogens.

Handover

Switching an on-going call to a different channel or cell in a wireless cellular network. Also known as "handoff."

Hard Drawn

Wire that has been drawn to its specific size and not annealed.

Hard Line

A Steel Pulling line.

Hardenability

A measure of the depth to which a specific ferrous alloy may be hardened by the formation of martensite upon quenching from a temperature above the upper critical temperature.

Hardness

The measure of some materials' resistance to deformation by surface indentation or by abrasion.

Harmonic

A sinusoidal component of the voltage that is a multiple of the fundimental wave frequency. Harmonics are primarily the result of the today's modern electronic equipment. Today's electronics are designed to draw current in "pulses" rather than in a smooth, sinusoidal manner as older, non-electronic equipment did. These pulses cause distorted current waveshapes, which in turn cause distortion of the voltage. Current and voltage harmonics can cause such problems as excessive heating of wiring, connections, motors, and transformers and can cause inadvertent tripping of circuit breakers.

Harmonic Content

The number of frequencies in the output waveform in addition to the primary frequency (50 or 60 Hz.). Energy in these harmonic frequencies is lost and may cause excessive heating of the load.

Harmonic Distortion

The presence of harmonics that change the AC voltage

waveform from a simple sinusoidal to complex waveform. Harmonic distortion can be generated by a load and fed back to the AC utility line, causing power problems to other equipment on the same circuit.

Harmonized Standard

A standard which has been drawn up by common agreement between national standards bodies notified to the European Commission by all member states and published under national procedures.

Hazard Risk Category

Categories defined by NFPA 70E-2004 to explain protection levels needed when performing tasks. The values range from 1 to 4. ATPV rated PPE is required for categories 1 through 4 as follows: 1- 4 cal/cm²; 2- 8 cal/cm²; 3- 25 cal/cm²; 4- 40 cal/cm².

H-Bridge

A circuit diagram which resembles the letter "H." The load is the horizontal line, connected between two pairs of intersecting lines. It is very common in DC motor-drive applications where switches are used in the "vertical" branches of the "H" to control the direction of current flow, and thus the rotational direction of the motor.

Head

The total vertical distance between the beginning of a hydro system diversion and the micro hydro turbine. The amount of power a turbine produces is proportional to the total available head.

Head Loss

Obstructions to the flow of water to a hydro turbine. Anything from the friction on the inside of the pipeline, to water turbulence in the pipe or fittings which change the pipeline direction can slow the water flow down, causing head loss.

Heat (electric)

The heat produced in a conductor by the passage of an electric current through it.

Heat Pump

Like an air conditioner or refrigerator, a heat pump moves heat from one location to another. In the cooling mode, heat pumps reduce indoor temperatures in the summer by transferring heat to the ground. Unlike an air conditioning unit, however, a heat pump's cycle is reversible. In winter, a heat pump can extract heat from the ground and transfer it inside. The energy value of the heat thus moved can be more than three times the cost of the electricity required to perform the transfer process.

Heat Rate
A measure of generating station thermal efficiency and generally expressed as Btu per net k/Wh. The heat rate is computed by dividing the total Btu content of the fuel burned (or of heat released from a nuclear reactor) by the resulting net kWh generated.

Heat Run Test
A test that is used to determine the increase in operating temperature at a given load.

Heat Sink
Mechanical device that is thermally-connected to a heat-producing electronic component, designed to conduct heat away from the device. Most heat sinks are aluminum and employ fins to increase surface area and encourage the transfer of heat to the ambient environment.

Heater
A heat source (gas or electric) used to adjust the temperature inside a dwelling from a cold to a warm condition.

Heating System
Energy Efficiency program promotion aimed at improving the efficiency of the heating delivery system, including replacement, in the residential, commercial, or industrial sectors.

Heavy Duty
A lightning impulse classifying current category for distribution class arresters defined by ANSI/IEEE C62.11. A heavy duty rated arrester has a 10,000 amperage impulse value crest (refer to normal duty).

Heavy Oil
The fuel oils remaining after the lighter oils have been distilled off during the refining process. Except for start-up and flame stabiliza-

tion, virtually all petroleum used in steam plants is heavy oil.

Hectare
A measure of large area. 1 hectare = 100 are = $10^4 \, m^2$

Hecto (h)
Decimal multiple prefix corresponding to a hundred or 10^2. This is not a preferred suffix.

Hedging Contracts
Contracts which establish future prices and quantities of electricity independent of the short-term market. Derivatives may be used for this purpose.

Helical
Wrapped in a spiral fashion. Refers to the way the strands of a conductor are laid.

Henry
The meter-kilogram-second unit of inductance, equal to the inductance of a circuit in which an electromotive force of one volt is produced by a current in the circuit which varies at the rate of one ampere per second.

Hertz (Hz)
An international measure of frequency or vibration equal to 1 cycle per second. The alternative current frequency used in North America is 60 hertz. In Europe and some other parts of the world it is 50 hertz.

Heterojunction
A region of electrical contact between two different materials.

Heuristic
A method of solving mathematical problems for which no algorithm exists. Involves the narrowing down of the field of search for a solution by inductive reasoning from past experience of similar problems.

Hybrid Vehicle, HEV
Hybrid vehicles (HEVs) have electric components, but use a fuel source (such as gasoline) to power the vehicle. The batteries can only be recharged by operating the vehicle (e.g., no plug). This is a growing market segment and many established auto manufacturers are producing hybrid vehicles today. The most conservative estimate for 2010 and beyond has J.D. Power forecasting a plateau of three percent hybrid pen-

etration in the U.S. market. The most optimistic and forward-looking prediction comes from Booz Allen Hamilton, a global strategy and technology-consulting firm. They predict that hybrid cars will make up 80 percent of the overall car market by 2015. (source for sales estimates: http://www.hybridcars.com/sales-numbers.html).

Hexadecimal
A number system consisting of 16 symbols, namely 0 to 9 and A to E.

Hexagonal Close-Packed (HCP)
A crystal structure found for some metals. The hcp unit cell is of hexagonal geometry and is generated by the stacking of close-packed planes of atoms.

High bit-rate Digital Subscriber Line (HDSL)
The oldest of the DSL technologies, it continues to be used by telephone companies deploying T1 lines at 1.5Mbps and requires two twisted pairs.

High Heat Value (HHV)
The high or gross heat content of the fuel with the heat of vaporization included; the water vapor is assumed to be in a liquid state.

High Intensity Discharge (HID) Lamp
An electric discharge lamp in which the light producing arc is stabilized by wall temperature and the arc tube has a bulb wall loading in excess of 3 watts per square centimeter. Examples of HID lamps include High Pressure Sodium, Metal Halide and Mercury Vapor.

High Level Data Link Control
An ITU-TSS link layer protocol standard for point-to-point and multi-point communications.

High Pass Filter
A filter designed to pass all frequencies above its cut-off frequency.

High Pot
A test done to confirm the reliability of an insulation system where a high voltage is applied.

High Pressure Sodium (HPS) Lamp
A High Intensity Dis-

charge light source in which the arc tube's primary internal element is Sodium Vapor. HPS is commonly used for roadway and area lighting.

High Rupturing Capacity Fuses (HRC Fuses)

The HRC fuse is usually a high-grade ceramic barrel containing the fuse element. The barrel is usually filled with sand, which helps to quench the resultant arc produced when the element melts. They are used for high current applications.

High Voltage

Voltage in a power line higher than the 110 to 220 volts used in most residences.

High Voltage Direct Current Transmission (Hvdc Transmission)

Power transmission carried out at high voltage direct current.

High Voltage Disconnect

The voltage at which a charge controller will disconnect the photovoltaic array from the batteries to prevent overcharging.

High Voltage System

An electric power system having a maximum roomean-square ac voltage above 72.5 kilovolts (kv).

High Voltage Test

A test which consists of the application of a specified voltage higher than the rated voltage between windings and frame, or between two or more windings, for the purpose of determining the adequacy of insulating materials and spacing against breakdown under normal conditions. [It is not the test of the conductor insulation of any one winding.

High-definition Television (HDTV)

An all-digital system for transmitting a TV signal with far greater resolution than the analog standards (PAL, NTSC, and SECAM). A high-definition television set can display several resolutions, (up to two million pixels versus a common television set's 360,000). HDTV offers other advantages such as greatly improved color encoding and the loss-free reproduction inherent in digital technologies.

High-Level Nuclear Waste

High-level radioactive waste is irradiated reactor fuel which includes the mass of uranium in the fuel assemblies and does not include the total weight of the fuel assemblies. Act 141, Section 10r(3), requires all Michigan electric suppliers, beginning January 1, 2002, to disclose to customers the environmental characteristics of the average fuel mix used to produce the electricity products purchased by the customers. This includes the average of the high-level nuclear waste generated in pounds per megawatt hour.

Highly accelerated stress test (HAST)

Highly accelerated steam and temperature

High-Pressure Mercury (Vapor) Lamp

Mercury vapor lamp, with or without a coating of phosphor, in which during operation the partial pressure of the vapor is of the order of 10^5 pa.

High-pressure Sodium Lamps

A sodium vapor in which the partial pressure of the vapor during operation is the order of 0.1 atmospheres.

High-Side

An element connected between the supply and the load. High-side current sensing applications measure current by looking at the voltage drop across a resistor placed between the supply and the load.

High-speed reclosing

A re-closing scheme where re-closure is carried out without any time delay other than required for deionization.

High-Speed Serial Interface (HSSI)

A short-distance communications standard for data rates from 2Mbps to 52Mbps.

High-Tech Troubleshooting

A procedure performed by a trained technician for the purpose of locating and identifying electrical problems within an electrical system.

Histogram

A type of graphical representation, used in statistics,

in which the frequency distribution is expressed by rectangles.

Hoistway

Any shaft way, hatchway, well hole, or other vertical opening or space in which an elevator or dumbwaiter is designed to operate.

Hole

The vacancy where an electron would normally exist in a solid; behaves like a positively charged particle.

Hole (electron)

For semi-conductors and insulators, a vacant electron state in the valence band that behaves as a positive charge carrier in an electric field.

Home Radio Frequency (RF)

Trademarked name for Home Radio Frequency, a networking technology which uses antennae and transmitters to provide wireless home networking via transmitted radio signals.

Home Run

In communications and structured wiring, when conductors are run from each device back to the source.

HomePlug

HomePlug (PowerLine) is an industry-standard method for transmitting data via the power lines. It can transmit audio, video, control signals, etc. Homeplug is a trademark of the HomePlug Powerline Alliance; Powerline is the generic term for the method.

Homogenous Materials

A homogeneous material is defined as either a raw material or a material applied during the construction of the product. For example, in reed blades plated with both Gold and a Ruthenium layer, the base metal (Nickel Iron alloy) and both layers are considered homogeneous materials and therefore must be considered separately.

Homojunction

The region between an n-layer and a p-layer in a single material photovoltaic cell.

Homopolymer

A polymer having a chain structure in which all mer units are of the same type.

Hooke's Law

Within the elastic limit, a strain is proportional to the stress producing it.

Hookstick

A hotstick that is used to operating switches and cutouts.

Horsepower

a measure of power (power is energy per time). For engines this is torque multiplied by rotational speed. In the automotive world, it is used to rate engines, but comparisons based on engine horsepower can be misleading because torque varies significantly with RPM for an ICE and the test conditions of the measurement must be carefully specified (e.g. SAE test procedures). In contrast, for electric motors, horsepower is simply defined as 746 Watts (0.746KW).

Horsepower Rated

A switch with a marked horsepower rating, intended for use in switching motor loads.

Hot

Current is present. A hot lead is the one carrying current along a circuit. It usually has black or red insulation. A hot circuit is one in which the breaker is closed and current is present.

Hot Arm

A device that is used to temporarily extend a conductor beyond the crossarm it was on.

Hot Cathode Lamp

An electrical discharge lamp whose mode of operation is that of an arc discharge.

Hot Spot

An undesirable phenomenon of PV device operation whereby one or more cells within a PV module or array act as a resistive load, resulting in local overheating or melting of the cell(s).

Hot Wire Instrument

An electrical instrument which depends upon the expansion (or change of resistance) of a wire which is heated by the passage of an electric current.

Hot Working

Any metal forming operation that is performed above a metal recrystallization temperature.

Hotstick

An insulated stick, ususlly made of fiberglass, that is used to work energized overhead conductors and operate electrical equipment that is overhead, underground and pad mounted.

Hot-Swap

A power supply line controller which allows circuit boards or other devices to be removed and replaced while the system remains powered up. Hotswap devices typically protect agaits overvoltage, undervoltage, and inrush current that can cause faults, errors, and hardware damage.

Hourly Metering

Tracking or recording a customer's consumption during specific periods of time that can be tied to the price of energy.

Hourly Non-Firm Transmission Service

Point-to-point transmission that is scheduled and paid for on an as-available basis and is subject to interruption.

House Service Meter

Energy meter at a consumer's premises, measuring power in kWh.

Human Body Model

An ESD test method where the ESD generator consists of a 100pF capacitor and a 1.5kohm series resistor.

Humidity

The humidity of the atmosphere is a measure of the water vapour present in the air.

Hybrid Electric Vehicle

A vehicle that combines conventional power production (e.g. an ICE) and an electric motor.

Hybrid System

A PV system that includes other sources of electricity generation, such as wind or diesel generators.
hydrogenated amorphous silicon-Amorphous silicon with a small amount of incorporated hydrogen. The hydrogen neutralizes dangling bonds in the amorphous silicon, allowing charge carriers to flow more freely.

Hydraulic Set Cement

A cement that sets through reaction with water.

Hydrocarbon

Organic compounds which contain only carbon and hydrogen.

Hydroelectric

An electric generating station in which a water wheel is driven by falling water, thus generating electricity.

Hydroelectric Generator

An electric generation system that is powered by falling water. Some hydroelectric facilities are operated as pumped storage facilities, where electricity is used to pump water uphill into a reservoir during times when demand is low and electricity is cheap to produce. During times when demand is high and electricity is more expensive to produce, the stored water is allowed to flow through hydroelectric generators. For some purposes, pumped storage facilities may not be categorized as hydroelectric generators.

Hydroelectric Plant

A plant in which the turbine/generators are driven by the kinetic energy of water. One common type of hydropower plant involves using a dam to store water in a reservoir and when released spins a turbine, creating electricity.

Hydrogenated Amorphous Silicon

Amorphous silicon with a small amount of incorporated hydrogen. The hydrogen neutralizes dangling bonds in the amorphous silicon, allowing charge carriers to flow more freely.

Hydrometer

A float type instrument used to determine the state-of-charge of a battery by measuring the specific gravity of the battery electrolyte (i.e., the amount of sulfuric acid in the electrolyte).

Hypoeutectoid Alloy

For an alloy system displaying a eutectoid, an alloy for which the concentration of solute is less than the eutectoid composition

Hysteresis

A physical phenomena wherein the path followed during relieving an applied stress lags that during applying the stress, so that on complete removal of the applied stress a strain remains. This particularly occurs in magnetic materials— the lagging of induced magnetism behind the magnetizing force.

Hysteresis Loss

Power loss in the magnetic core due to hysteresis.

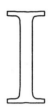

I/Q

1. I/Q modulation is a method for combining two channels of information into one signal so that they can be separated at a later stage. Two quadrature carriers, 90 degrees out of phase, are modulated, then combined. Abbreviated from "in-phase/quadrature-phase" which refers to the two carrier signals' phase relationship.

2. I_Q (Q should be subscripted but sometimes printed as "IQ" without subscripting): Quiescent current: The current consumed when a circuit is in a quiet state, driving no load and if appropriate, with its inputs not cycling.

3. Intelligence quotient, a measure in which electrical engineers invariably excel.

Ideal Current Source

A source which maintains the source current at a predefined value independent of the load conditions.

Ideal Dependent Source

An active element in which the source voltage or current is controlled precisely by another voltage or current.

Ideal Operational Amplifier (Op Amp)

An operational amplifier with infinite open-loop gain, infinite input impedance and zero output impedance.

Ideal Source

An ideal independent source is an active element that provides a specified voltage or current that is completely independent of the remaining circuit elements.

Ideal Voltage Source

A source which maintains the source voltage at a predefined value independent of the load conditions. In other words the terminal voltage is maintained equal to the internal emf.

Ideality Factor
A constant adjustment factor used to correct for discrepancies between an ideal PN junction equation and a measured device.

III-V (Three-Five) Materials
Elemental materials that occupy groups III and V of the Periodic Table of the Elements.

Illuminance
The density of the luminous flux incident on a surface. It is the quotient of the luminous flux multiplied by the area of the surface when the later is uniformly illuminated.

Image Frequency
Receivers typically convert RF signals to a lower Intermediate Frequency (IF) for demodulation. In addition to the IF, a second signal, called the "image frequency" is often generated and filtered out.

Image Rejection
The measure of a receiver's ability to reject signals at its image frequency. It is normally expressed as the ratio, in dB, of the receiver's sensitivity at the desired frequency versus the sensitivity at the image frequency.

Imaginary
Numbers with negative squares. S-1 is the base of such numbers.

Imaginary Axis
The y-axis in the complex plane.

Imaginary Operator J
An multiplier or operator with a magnitude of unity and an anticlockwise rotation of 90°. It also has the value S-1 in the complex domain.

Impact Energy
A measure of the energy absorbed during the fracture of a specimen of standard dimensions and geometry when subjected to very rapid (impact) loading. Charpy and izod impact tests are used to measure this parameter, which is important in assessing the ductile-to-brittle transition behavior of a material.

Impedance

Impedance is the opposition offered by a material to the flow of an electrical current and is a characteristic of AC systems. Impedance has two parts—resistance and reactance. Reactance has two components, capacitive reactance and inductive reactance. The properties of these last two components are dependent upon the frequency.

Impregnation

The process of filling the pores of paper and similar material in order to improve its insulation properties.

Impulse

A disturbance of the voltage waveform that is less than about one millisecond. Voltages can rise to hundreds or even thousands of volt in a very short period of time. An impulse may be additive or subtractive.

Impulse Function

A mathematical function with zero magnitude other than at zero time, where it has an infinite magnitude. The magnitude of an impulse function is defined as its time integral.

Impulse Generator

In most impulse generators, certain capacitors are charged in parallel through high series resistances, and then discharged through a combination of resistors and capacitors, giving rise to the required surge waveform (usually double exponential) across the test device.

Impulse Response

Behaviour of a circuit when the excitation is the unit impulse function. The excitation function may be a voltage or a current.

Impulse Test

·Tests to confirm that the insulation level is sufficient to withstand overvoltages, such as those caused by lightning strikes and switching.

Impulse Turbine

Impulse turbines produce power when a jet of water from an enclosed diversion pipeline 'shoots' through a small nozzle directly onto the turbine runner. Impulse turbines are best for 'high head' sites (with 20 feet of head or more), but they do not require very high flow rates. Pelton and Turgo turbines are two of the most common impulse turbine families.

In Sight From

(within sight from, within sight) Where this Code specifies that one equipment shall be "in sight from", "within sight from" or m "within sight", etc. of another equipment, the specified equipment is to be visible and not more that 50? distant from the other

Incandescent

Designed for use with all manufactured incandescent lamps, most of which have threaded bases.

Incandescent (Electric) Lamp

Lamp in which light is produced by means of an element heated to incandescence by the passage of an electric current.

Incandescent Light Bulbs

Incandescent light bulbs produce light by passing electricity through a thin filament, which becomes hot and glows brightly. Incandescent light bulbs are less energy-efficient than fluorescent lamps, because much of the electrical energy is converted to heat instead of light. The heat produced by these bulbs not only wastes energy, but can also make a building's air conditioning system work harder and consume more energy.

Incentive

A rebate or some form of payment used to encourage people to implement a given demand-side management (DSM) technology. The incentive is calculated as the amount of the technology costs that must be paid by the utility for the participant test to equal one and achieve the desired benefit/cost ratio to drive the market.

Inch

A measure of length in the imperial system. It is now defined as follows. 1 inch = 25.400 mm

Incidence Matrix

A connection matrix having elements 1, -1 or 0 dependent on whether a particular connection is present and having the same sign as the reference, has the opposite sign to the reference or not connected at all.

Incident Energy

The amount of energy impressed on a surface, a certain distance from the source, generated during an electrical arc event. Often measured in calories per centimeter squared. (cal/cm²)

Incident Light

Light that shines onto the face of a solar cell or module.

Incidental Light Traffic

Refers to a grade level Reinforced Polymer Concrete or Fiberglass Reinforced Plastic box or Cover load rating of 10,400lbs. This rating is derived from incidental single vehicle tire contact estimated at a maximum of 8000lbs with an impact factor of 30% added (8000 x 1.30). In the US, this rating is also referred to as, "Light Traffic", 10K or "Parkway". Application is limited generally behind a curb or guardrail, in sidewalks, or in other locations where vehicle traffic is not intended.

Incremental Effects

The annual effects in energy use (measured in megawatthours) and peak load (measured in kilowatts) caused by new participants in existing DSM programs and all participants in new DSM programs during a given year. Reported Incremental Effects should be annualized to indicate the program effects that would have occurred had these participants been initiated into the program on January 1 of the given year. Incremental effects are not simply the Annual Effects of a given year minus the Annual Effects of the prior year, since these net effects would fail to account for program attrition, degradation, demolition, and participant dropouts.

Independent Power Producer (IPP)

Private entrepreneurs who develop, own or operate electric power plants fueled by alternative energy sources such as biomass, cogeneration, small hydro, waste-energy and wind facilities. Organisations with generating capacity that are not associated with traditional electricity utilities.

Independent System Operator (ISO)

An entity which handles

access to the transmission system to ensure that suppliers get fair access to use it, and that consumers receive electricity reliably. ISOs are designed to ensure that transmission-owning utilities cannot use their own transmission resources to gain a generation price advantage over other power suppliers.

Independent Time Measuring Relay

A measuring relay, the specified time for which can be considered as being independent, within specific limits, of the value of the characteristic quantity.

Index

The number indicating the power to which the quantity is raised.

Indicating Instrument

A piece of equipment in which the output is given as the deflection of a needle or the reading of a counter.

Indirect Contact

Contact of persons or livestock with exposed conductive parts which have become live under fault conditions.

Indirect Lighting

Lighting by means of luminaires with a light distribution such that not more than 10 per cent of the emitted luminous flux reaches the working plane direct, assuming that this plane is unbounded.

Indirect Utility Cost

A utility cost that may not be meaningfully identified with any particular DSM program category. Indirect costs could be attributable to one of several accounting cost categories (i.e., Administrative, Marketing, Monitoring & Evaluation, Utility-Earned Incentives, Other). Accounting costs that are known DSM program costs should not be reported under Indirect Utility Cost, rather those costs should be reported as Direct Utility Costs under the appropriate DSM program category.

Indium Oxide

A wide band gap semiconductor that can be heavily doped with tin to make a highly conductive, transparent thin film. Often used as a front contact or one component of a heterojunction solar cell.

Induced Current

Current in a conductor resulting from the application of a time varying electromagnetic field.

Induced Voltage

A voltage produced around a closed path or circuit by a change of magnetic flux linking that path.

Inductance

The voltage across an inductor is directly proportional to the rate of change of the current through it divided by the rate of change of time (difference current/difference time = di/dt). The proportionality constant which makes this true is L, the inductance of the inductor component. It is denoted by L and its units are the Henry (H). Therefore, the voltage v across an inductor is given by $v = L*(di/dt)$.

Induction

Induction is the process by which charge is moved in a conductor by the presence of an electric field. In wires this will lead to a current, in discrete conducting objects it will lead to local charging—ie. the side near the inducing charge will become the opposite charge and the far side will acquire a similar charge leaving the overall charge of the object unchanged.

Induction Heating

A form of heating in which electrically conducting material is heated as a result of the electric currents induced in it by an alternating magnetic field.

Induction Machine

Induction machines run at a speed slightly different to synchronous speed as the difference speed, known as slip, is required to generate torque. Induction motors run at sub-synchronous speed where as induction generators run at super-synchronous speed.

Induction Meter

The induction meter depends on the torque produced by the reaction between a flux (whose value depends on the value of the current in one coil) and the eddy currents which are induced in a non-magnetic disc (usually aluminium) by another flux (produced by current in a second coil). Since the action depends on induction, they can be used to measure alternating quantities only. The meter would have a deflection proportional to the product of the two currents.

Inductive Coupling

The coupling or linkage of two circuits by the changing magnetic lines of force.

Inductive Kickback

The very rapid change in voltage across an inductor when current flow is interrupted. Snubber diodes are often used to channel this energy in relays, and other inductive loads. Kickback can be a problem (causing EMI and component failure); or it can be used in power supply circuits to develop higher or opposite-polarity voltages from a single supply.

Inductor

A circuit element which is a wire wound into a coil to create a magnetic field.

Inductor (L)

A coil of wire which has the property of inductance.

Industrial

The industrial sector is generally defined as manufac-turing, construction, mining, agriculture, fishing, and forestry establishments (Standard Industrial Classi-fication [SIC] codes 01-39). The utility may classify industrial service using the SIC codes, or based on demand or annual usage exceeding some specified limit. The limit may be set by the utility based on the rate schedule of the utility.

Industrial, Scientific and Medical (ISM)

Radio frequency bands made available for use by communication equipment without license, within certain maximum emitted power limits. Equipment which uses the ISM band must tolerate interference from other such equipment. Common uses include WiFi (802.11a, b, and g) and cordless phones.

Inertia

Tendency of a body to preserve its state of rest or uniform motion.

InfiniBand

InfiniBand architecture is an industry standard, channel-based, switched-fabric, interconnect architecture for servers. InfiniBand architecture changes the way servers are built, deployed, and managed.

Infinitesimal

A very small quantity tending to zero, without actually being zero.

Infinity

A very large quantity greater than any assignable quantity and tending to the inverse of zero.

Influence Quantity

A quantity which is not the subject of the measurement but which influences the value of the output signal for a constant value of the measured.

Infrared (IR)

Light that has a frequency below the visible light spectrum, used for remote controls, line-of-sight wireless data, and night vision applications, among others.

Infrared Cameras

Energy contractors use infrared cameras to look at the heat leaking into or out of your house. The infrared camera "sees" the heat and can show "hot spots" where a lot of heat is being lost. This helps to identify the places where your home's energy efficiency can be improved.

Infrared Radiation

Electromagnetic radiation whose wavelengths lie in the range from 0.75 micrometer to 1000 micrometers; invisible long wavelength radiation (heat) capable of producing a thermal or photovoltaic effect, though less effective than visible light.

Inorganic

Not belonging to the large class of carbon compounds which are termed organic.

Input

The intake or energy absorbed by a machine during its operation, as distinguished from the output of useful energy delivered by it.

Input Back-Off (IBO)

In a power amplifier, a measure of how far you must reduce the input power in order to receive the desired output linearity and power. Stated differently, the ratio between the input power that delivers maximum power to the input power that delivers the desired linearity.

Input Common-Mode Voltage Range (CMVR) (V)

Common-mode voltage range (CMVR) or Input Voltage Range (IVR): For signal processing devices with differential inputs, such as an op amp, CMVR is the range of common mode signal for which the amplifier's operation remains linear.

If we let the voltage present on the "-" input equal V1, and the voltage on the "+" input equal V2, then the common mode voltage is VCM = (V1+V2)/2.

Some op amps, for instance, will only allow the common mode voltage of a signal to come within a diode drop or so of the power supply rails. Many of Maxim's op amps will allow the common mode input voltage to go all the way to one or both supply rails. Some even allow inputs beyond the supply rails (Beyond-The-Rails™).

Input Power Frequency

This is the frequency range that can be input into the suppressor without damaging it.

Input Quantity

The quantity, or one of the quantities, which constitute the signals received by the transducer form the measured system.

Input Voltage

This is determined by the total power required by the alternating current loads and the voltage of any direct current loads. Generally, the larger the load, the higher the inverter input voltage. This keeps the current at levels where switches and other components are readily available.

Inrush Current

A momentary input current surge, measured during the initial turn-on of the power supply. This current reduces to a lower steady-state current once the input capacitors charge. Hotswap controllers or other forms of protection are often used to limit inrush current, because uncontrolled inrush can damage components, lower the available supply voltage to other circuits, and cause system errors.

Insertion Force

The effort, usually measured

in ounces, required to engage mating components.

Insertion Loss

This is the loss that occurs as signals pass through a passive device. Insertion loss occurs in all devices which do not amplify the signal. Also called "feed through loss".

Insolation

Sunlight, direct or diffuse; from 'incident solar radiation.' Not to be confused with 'insulation.' Equal to about 1000 watts per square meter at high noon in Dodge City.

Installation

An electrical installation is a combination of electrical equipment installed to fulfil a specific purpose and having coordinated characteristics.

Installed capacity

The total generating units' capacities in a power plant or on a total utility system. The capacity can be based on the nameplate rating or the net dependable capacity.

Instantaneous Frequency

The rate of change of phase angle (in rad/s) or additionally divided by 2? (in Hz).

Instantaneous Relay

A relay that operates and resets with no intentional time delay.

Instantaneous Value

The value of an alternating current or voltage at any specified instant in a cycle.

Instrinsic Semiconductor

A semiconductor material for which the electrical behavior is characteristic of the pure material.

Instructed Person

A person adequately advised or supervised by skilled persons to enable him/her to avoid dangers which electricity may create.

Instrument Transformer

A transformer that is only designed to reduce current or voltage from a primary value that is too pass directly through a meter or instrument, to a proportional low level that can safely be applied.

Insulated Gate Bipolar Transistor

A special design of transistor that is suitable for handling high voltages and currents. Often used in static power control equipment such as inverters, or controlled rectifiers, due to the flexibility of control of the output.

Insulation

1) A non-conductive material used on a conductor to separate conducting materials in a circuit. 2) The non-conductive material used in the manufacture of insulated cables. South wire Company manufacturers insulated cables. Speed Systems manufacturers tools to strip insulation from cable.

Insulation Class

A letter or number that designates the temperature rating of an insulation material or system with respect to thermal endurance.

Insulation Co-Ordination

Insulation co-ordination now comprises the selection of the electric strength of the various equipment in relation to the voltages which can appear on the system for which the equipment is intended. The overall aim is to reduce to an economically and operationally acceptable level the cost and disturbance caused by insulation failure and resulting system outages.

Insulation Failure

Fault between the phase conductor and non-current carrying metallic parts of an electrical equipment, as a result of which high voltages may appear on the frames of equipment and may be dangerous to a person coming in contact with it.

Insulation Level

It defines the level of insulation with regard to power frequency and with regard to surges. For equipment rated at less than 300 kV, it is a statement of the Lightning impulse withstand voltage and the short duration power frequency withstand voltage. For equipment rated at greater than 300 kV, it is a statement of the Switching impulse withstand voltage and the power frequency withstand voltage.

Insulation Resistance

The electrical resistance measured between insulated terminals.

Insulators

Material containing seven or eight valence electrons are known as insulators. Insulators are materials that resist the flow of electricity. When the valence shell of an atom is full, the electrons are held tightly and are not given up easily. Some good examples of insulator materials are rubber, plastic, glass, and wood. The energy of the moving electron is divided so many times that it has little effect on the atom. Any atom that has seven or eight valence electrons is extremely stable and does not easily give up an electron.

Intake

The point at which water is diverted from a river or stream to the turbine via a diversion. A trash rack/filter and settling tank are often installed at the intake point to prevent debris and sand or silt from reaching the turbine.

Intangible Transition Charge

The amounts on all customer bills, collected by the electric utility to recover transition bond expenses.

Integer

A whole number.

Integral Nonlinearity

A measure of a data converter's ability to adhere to an ideal slope in its transfer function. It can be specified using end-point or best-straight-line fit. Each of these approaches can yield very different numbers for the same data converter.

Integrated Circuit (IC)

1. A semiconductor device that combines multiple transistors and other components and interconnects on a single piece of semiconductor material.
2. Internally Connected

Integrated Heat Spreader

An Integrated Heat Spreader (IHS) is the surface used to make contact between a heatsink or other thermal solution and a CPU or GPU processor.

Integrated Resource Plan (IRP)

A comprehensive and systematic blueprint developed by a supplier, distributor, or end-user of energy who has evaluated demand-side and supply-side resource options and economic parameters and determined which options will best help them meet their energy goals at the lowest reasonable energy, environmental, and societal cost.

Integrating Meter

A meter whose output is proportional to the integrated value of a quantity over time. They are usually with rotating discs where the revolutions correspond to the time of integration.

Integrator

An op amp whose output is proportional to the integral of the input signal.

Intel Mobile Voltage Positioning (IMVP)

A technology in which the processor voltage (VCC) is dynamically adjusted, based on the processor activity, to reduce processor power. It allows higher processor clock speed at a given power consumption; or lower consumption at a given clock frequency.

Intelligent Electronic Device

Equipment containing a microprocessor and software used to implement one or more functions in relation to an item of electrical equipment. IED is a generic term used to describe any microprocessor-based equipment, apart from a computer.

Intensity (Lighting)

The brightness of light in a given direction. Luminous intensity may be expressed in Candelas (cd) or in Lumins.

Interchange (Electric utility)

The agreement among interconnected utilities under which they buy, sell and exchange power among themselves. This can, for example, provide for economy energy and emergency power supplies.

Interchangeable

A receptacle or combination of receptacles with a common mounting dimension that may be installed on a single or multiple-opening mounting strap.

Interconnect

A conductor within a module or other means of connection which provides an electrical interconnection between the solar cells.

Interconnection

The linkage of transmission lines between two utility, enabling power to be moved in either direction. Interconnections allow the utilities to help contain costs while enhancing system reliability.

Interconnection Voltage

The nominal voltage at which the grid interconnection is made.

Interdepartmental Service (Electric)

Interdepartmental service includes amounts charged by the electric department at tariff or other specified rates for electricity supplied by it to other utility departments.

Interfacial Seal

Sealing of a two-piece, multiple contact connector over the whole area of the interface to provide sealing around each contact.

Interleave

To organize the data sectors on a computer hard disk, so the read/write heads can access information faster.

Interlock

A device connected in such a way that the motion of one part is held back by another part.

Intermediate (Lighting)

Those areas of a municipality often characterized by moderately heavy nighttime pedestrian activity such as in blocks having libraries, community recreation centers, large apartment buildings, industrial buildings or neighborhood retail stores.

Intermediate Class Arrester

Surge arresters with a high energy handling capability. These are generally voltage classed at 3-120kV.

Intermediate Frequency (IF)

Radio communications systems modulate a carrier frequency with a baseband signal in order to achieve radio transmission. In many cases, the carrier is not modulated directly. Instead, a lower IF signal is modulated and processed. At a later circuit stage, the IF signal is converted up to the transmission frequency band.

Intermediate Load (Electric System)

The range from base load to a point between base load and peak. This point may be the midpoint, a percent of the peakload, or the load over a specified time period.

Intermetallic

A compound of two metals that has a distinct chemical formula. The bonds in intermetallic compounds are often partly ionic.

Intermittent Resources

Resources whose output depends on some other factory that cannot be controlled by the utility e.g. wind or sun. Thus, the capacity varies by day and by hour.

Intermodulation

A process whereby signals mix together in a circuit and nonlinearities in the circuit create undesired output frequencies that are not present at the input.

Intermodulation Distortion

An RF signal defect in which non-linear circuits or devices create new frequency components not in the original signal, including the common harmonic and two-tone distortion effects.

Internal Combustion Engine

An engine that burns fuel inside a reaction chamber to create pressure inside the chamber that is converted into rotary motion. ICE engines are typically based on the Otto cycle, Atkinson cycle, or Wankel engine.

Internal Combustion Plant

A plant in which the prime mover is an internal combustion engine. An internal combustion engine has one or more cylinders in which the process of combustion takes place, converting energy released from the rapid burning of a fuel-air mixture into mechanical energy. Diesel or gas-fired engines are the principal types used in electric plants. The plant is usually operated during periods of high demand for electricity.

Internal Discharge

A discharge occurring within a material.

Internal Impedance (Battery)

The opposition to the flow of alternating current at a particular frequency in a cell or battery at a specific state-of-charge and temperature.

Internal Reference (Int. Ref.)

An on-chip voltage reference.

Internal Resistance
The resistance to the flow of an electric current within the cell or battery.

Internal Resistance (Battery)
The opposition or resistance to the flow of Direct Electric Current within a cell or battery; The sum of the ionic and electronic resistance of the cell components. Its value may vary with the current, state-of-charge, temperature, and age. With an extremely heavy load, such as an engine starter, the cell voltage may drop to approximately 1.6. This voltage drop is due to the internal resistance of the cell. A cell that is partly discharged has a higher resistance than a fully charged cell, hence it will have a greater voltage drop under the same load. This internal resistance is due to the accumulation of lead sulfate on the plates. The lead sulfate reduces the amount of active material exposed to the electrolyte, hence it deters the chemical action and interferes with the current flow.

International Candle
[An old unit of luminous intensity, now replaced by the candela.] A point source emitting light uniformly in all directions at one-tenth the rate of the Harcourt pentane lamp burning under specific conditions.

International System of Units (SI)
A universal system of units in which the following six units of measure are considered basic: meter, kilogram, second, ampere, Kelvin degree and acndela.

Internet Protocol Address (IP Address)
This address is a 32 bit, unique string of numbers that identifies a computer, a printer, or another device on the internet. The IP address consists of a quartet of numbers separated by periods.

Interpolation
A process of filling in intermediate values or terms between known values or terms.

Interrupter
An element designed to interrupt specific currents under specified conditions.

Interrupter Rating
The highest current at rated

voltage that a device is intended to interrupt under standard test conditions.

Interrupter Switch

A switch equipped with an enterrupter for making or breaking connections under load

Interruptible Gas

Gas sold to customers with a provision that permits curtailment or cessation of service at the discretion of the distributing company under certain circumstances, as specified in the service contract.

Interruptible Load

Refers to program activities that, in accordance with contractual arrangements, can interrupt consumer load at times of seasonal peak load by direct control of the utility system operator or by action of the consumer at the direct request of the system operator. It usually involves commercial and industrial consumers. In some instances the load reduction may be affected by direct action of the system operator (remote tripping) after notice to the consumer in accordance with contractual provisions. For example, loads that can be interrupted to fulfill planning or operation reserve requirements should be reported as Interruptible Load. Interruptible Load as defined here excludes Direct Load Control and Other Load Management. (Interruptible Load, as reported here, is synonymous with Interruptible Demand reported to the North American Electric Reliability Council on the voluntary Form EIA-411, "Coordinated Regional Bulk Power Supply Program Report," with the exception that annual peakload effects are reported on the Form EIA-861 and seasonal (i.e., summer and winter) peakload effects are reported on the EIA-411).

Interruptible Power

This refers to power whose delivery can be curtailed by the supplier, usually under some sort of agreement by the parties involved.

Interruptible Rate

Tariff rate for the provision of power at a lower rate to large industrial and commercial consumers who agree to reduce their electricity use in times of peak demand.

Interrupting Medium

The "fluid" used to interrupt the flow of electric current in a switch or circuit breaker. In high power equipment, this may be oil, insulating gas or even no material, as is the case of vacuum interruption which works because in a vacuum there is no material that can be ionized to perpetuate the flow of electric current when the contact par and an arc tries to form. Air also can be used as an interrupting medium where the design of the interrupter is such that a stream of air blows-out and cools the arc path.

Interrupting Rating

The rating of a device to interrupt the flow of power or current, generally applied to a circuit breaker or a switch.

Interruption

The loss of electric service to one or more customers or other facilities. It is the result of one or more component outages.

Interruption Duration

The period from the initia-

tion of an interruption to a customer or other facility to the time the service is restored.

Interruption, Momentary

An interruption of a duration limited to the period required to restore service by automatic or supervisory controlled switching operations or by manual switching at locations where operators are immediately available. Such switching operations must be completed within five (5) minutes, including all reclosing operations.

Interruption, Sustained

Any interruption not classified as momentary. Any interruption longer that five (5) minutes.

Interstices

The space between two or more objects, such as the individual strands in a stranded conductor or conductors in a cable.

Inter-Symbol Interference (ISI)

A form of interference that occurs when echoes of a radio-signal interfere with the original signal. ISI can reduce the effective data rate of wire-less LAN transceivers.

Interval Metering

The process by which power consumption is measured at regular intervals in order that specific load usage for a set period of time can be determined.

Intranet

A restricted access network that works like the internet but is not. Usually owned and managed by a corpora-tion, an Intranet enables a company to share its re-sources with its employees without confidential infor-mation being made available to everyone with internet access.

Intrinsic

A word used to indicate a semiconductor that has no impurities.

Intrinsic Error

An error determined when the transducer is under ref-erence conditions.

Intrinsic Layer

A layer of semiconductor material, used in a photovol-taic device, whose properties are essentially those of the

pure, undoped, material.

Intrinsic Semiconductor

An undoped semiconduc-tor.

Intrinsically Safe

A term used to define a level of safety associated with the electrical controls or cir-cuits.

Intrinsically Safe Device

A device, instrument or component that will not pro-duce any spark or thermal effects under any conditions that are normal or abnormal that will ignite a specified gas mixture.

Invariant Point

A point on a binary phase diagram at which three phases are in equilibrium.

Inverse Square Law

A law which states that the intensity of an effect due to a source varies inversely as the square of the distance between.

Inverse Time Delay Relay

A dependent time delay re-lay having an operating time which is an inverse function of the electrical characteris-tic quantity.

Inverter

An electrical device which is designed to convert direct current into alternating current. This was originally done with rotating machines which produced true sine wave ac output. More recently this conversion has been performed more economically and efficiently using solid state electronics. However, except for the most expensive models, these devices usually do not produce perfect sine wave output. This sometimes can result in electromagnetic interference with other sensitive electronic devices.

Inverting Amplifier

Reverses the polarity of the input signal while amplifying it.

Inverting Switching Regulator

A switch-mode voltage regulator in which output voltage is negative with respect to its input voltage.

Investor-Owned Utility (IOU)

A stockholder-owned utility company that provides public utility services to retail customers for a profit. In Michigan, investor-owned electric utilities are regulated by the MPSC.

Ion

An electrically charged atom or group of atoms that has lost or gained electrons; a loss makes the resulting particle positively charged; a gain makes the particle negatively charged.

Ionisation

Ionisation is the process by which an electron is removed or added to an atom, leaving the atom with a nett negative charge (negative ion) or a nett positive charge (positive ion). A process of formation of ions.

Irradiance

The direct, diffuse, and reflected solar radiation that strikes a surface. Usually expressed in kilowatts per square meter. Irradiance multiplied by time equals insolation.

irreversible process

A process that is not fully reversible.

Islanded Operation

The situation that arises when a part of the electrical system is disconnected from the main grid and is energised by one or more generators connected to it.

Islanding

The process whereby a power system is separated into two or more parts, with generators supplying loads connected to some of the separated systems.

Isokeraunic Level

Contours of constant keraunic level or thunder days.

Isolated Ground

Receptacles intended for use in an Isolated Grounding system where the ground path is isolated from the facility grounding system. The grounding connection on these receptacles is isolated from the mounting strap.

Isolation

The degree to which a device can separate the electrical environment of its input from its output, while allowing the desired transmission to pass across the separation. A function intended to cut off for reasons of safety the supply from all, or a discrete section, of the installation by separating the in-

stallation or section from every source of electrical energy.

Isolation diode

A diode which prevents one segment of a PV array from interacting with another array segment. Usually used to prevent array energy from flowing backwards through a sub-voltage series string. May also serve the function of blocking diode.

Isolation Link

A metal link used in series with a fusing device that melts and prevents refusing/re-enerization of a transformer.

Isolation Transformer

A transformer with physically separate primary and secondary windings. An isolation transformer does not transfer unwanted noise and transients from the input circuit to the output windings.

Isolator

A mechanical switching device which, in the open position, complies with the requirements specified for isolation. An isolator is otherwise known as a disconnector.

Isotactic

A type of polymer chain configuration wherein all side groups are positioned on the same side of the chain molecule.

Isothermal

A process that is kept at a constant temperature.

Isotropic

Substances exhibiting uniform properties throughout, in all directions.

Isotropic Antenna

A theoretical, ideal antenna having a signal range of 360 degrees. It is used as a baseline for measuring a real antenna's strength signal, in dBi, where i represents Isotropic Antenna.

Itemized Service

Charges for electric service broken down into its basic components, including generation, transmission, distribution, billing and metering services.

I-Type Semiconductor

Semiconductor material that is left intrinsic, or undoped so that the concentration of charge carriers is characteristic of the material itself rather than of added impurities.

I-V Curve

A graphical presentation of the current versus the voltage from a photovoltaic device as the load is increased from the short circuit (no load) condition to the open circuit (maximum voltage) condition. The shape of the curve characterized cell performance.

I-V Data

The relationship between current and voltage of a photovoltaic device in the power-producing quadrant, as a set of ordered pairs of current and voltage readings in a table, or as a curve plotted in a suitable coordinate system (i.e., Cartesian).

Izod Impact Test

One of two tests that may be used to measure the impact energy of standard notched specimen.

J
Joule(s)

Jacket
A covering over insulated conductors for the purpose of electrical, chemical, and physical protection.

Jacobi's Law
A law of electric motors which states that the maximum work of a motor is performed when its counter electromotive force is equal to one half the electromotive force expended on the motor.

Jamming
When the combined diameters of three cables roughly equal the interior diameter of the conduit, the cables can line up linearly as they are pulled around the bend. The cables then wedge against the conduit wall as they are forced towards the inside of the bend. The wedged (or jammed) cables are "stuck." To pull jammed cables with enough force to get them through a bend usually ruins the cable by ripping off the jacket or crushing the insulation.

Jitter
The slight movement of a transmission signal in time or phase that can introduce errors and loss of synchronization. More jitter will be encountered with longer cables, cables with higher attenuation, and signals at higher data rates. Also, called phase jitter, timing distortion, or intersymbol interference.

Joint
The tying together of two single wire conductors so that the union will be good, both mechanically and electrically.

Joule

A measure of the amount of energy delivered by one watt of power in one second, or 1 million watts of power in one microsecond. The joule rating of a surge protection device is the amount of energy that it can absorb before it becomes damaged. In comparing surge protection performance, the Joule rating of a surge suppressor is less important than the let-through voltage rating. This reflects the fact that surge suppressors may protect equipment by deflecting surges as well as absorbing them. There is no standard for measuring the joule rating of surge suppressors which has resulted in wildly exaggerated claims by unscrupulous vendors.

Joule Rating

The measurement of a Surge Protective Device's ability to absorb heat energy created by transient surges. Note that the Joule rating is not a part of IEEE or UL Standards. It is not as significant a specification as Clamping Voltage, Maximum Surge Current and other parameters recognized by these agencies.

Joule's Law

Defines the relationship between current in a wire and the thermal energy produced. In 1841an English physicist James P. Joule experimentally showed that $W = I^2 \times R \times t$ where I is the current in the wire in amperes, R is the resistance of the wire in Ohms, t is the length of time that the current flows in seconds, and W is the energy produced in Joules.

Joule's Equivalent

Mechanical equivalent of heat. 4.185 J/cal

Jumper

An electrical connection between two points.

Junction

A region of transition between semiconductor layers, such as a p/n junction, which goes from a region that has a high concentration of acceptors (p-type) to one that has a high concentration of donors (n-type).

Junction Box

A box containing splices in cables. Has a removable cover that must be accessible (cannot be buried in ceilings and walls). Also called a *J-box*.

Junction Diode

A semiconductor device with a junction and a built-in potential that passes current better in one direction than the other. All solar cells are junction diodes.

Junction Diode Sensor

The use of a PN junction on a silicon die for determining die temperature.

Junction, 200 Amp

A "200 Amp Junction" is a molded synthetic and composite device used to connect two or more 200 Amp rated cables operating at Medium Voltage (4-35KV nominal). Connections to the cables are made via "200 Amp Elbows".

Junction, 600 Amp

A "600 Amp Junction" is a molded synthetic and composite device used to connect two or more 600 or 900 Amp rated cables operating at Medium Voltage (4-35KV nominal). Connections to the cables are made via "T-Bodies".

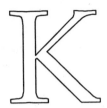

k

1. Kilo: Metric unit representing 1000. E.g.: 1kHz is a 1 kilohertz (1000 Hertz). Note that the k is always lowercase.

2. Kelvin: Temperature scale. Zero K is defined as absolute zero. 273.15K is 0 degrees C.

Note that temperatures on the kelvin scale are called kelvins, not "degrees kelvin." The K symbol is uppercase and used without a degree symbol. The word "kelvin" in this context is not capitalized.

Karnaugh Map

An arrangement of cells representing the combinations of variables in a Boolean expression and used for a systematic simplification of the expression.

K-Bus (Courier)

Term used for the courier protocol on K-Bus interface for K-Relay range manufactured by Alstom.

kcmil

One thousand circular-mils.

Keep-Out Zone

The area on or near a CPU or GPU processor that the circuit board layout design can not use, due to thermal management components, cooling, and mounting constraints.

Kelvin (K)

The kelvin is unit of thermodynamic temperature. It is a fundamental unit. It is defined as the fraction 1/273.16 of the thermodynamic temperature of the triple point of water [1967]

Kelvin Double Bridge

It is a double bridge arrangement, which is an extension of the Wheatstone bridge, for the precise measurement of low resistance. The errors due to contact and lead resistances are eliminated by the additional bridge incorporated.

Keraunic Level

Number of days in the year in which thunder is heard.

Key

A lampholder with a flat or round "key" knob that operates an internal switching mechanism ("Keyless" lampholders do not provide an internal switching mechanism).

Kibi (Ki)

Binary multiple prefix corresponding to kilobinary or 2^{10} or $(2^{10})^1$ or 1024. [IEC 1998]

Kilo (K)

A metric prefix meaning 1000 or 10^3.

Kilo Watt Hour (kWh)

Kilo Watt Hour, the use of one thousand watts for one hour.

Kilovolt (kv)

1,000 volts. The amount of electric force carried through a high-voltage

transmission line is measured in kilovolts.

Kilovolt Ampere (kVA)

1) Apparent Power expressed in Thousand Volt-Amps. 2) Kilovolt Ampere rating designates the output which a transformer can deliver at rated voltage and frequency without exceeding a specified temperature rise.

Kilowatt

A unit of electrical power, equal to one thousands watts. Electric power is usually expressed in kilowatts. As the watt is equal to 1/746 horsepower, the kilowatt or 1,000 watts = 1.34 hp. Careful distinction should be made between kilowatts and kilovolt amperes.

Kinetic Energy

Energy due to motion. Wind and moving water have kinetic energy, which is converted to electrical energy by a wind or hydro turbine.

Kirchoff's Current Law (KCL)

States that the algebraic sum of the currents entering a node (or a closed boundary) is zero.

Kirchoff's Voltage Law (KVL)
States that the algebraic sum of the voltages around a closed loop is zero.

Klydonograph
An instrument for the detection and recording of the occurrence of lightning in transmission lines.

Knee-Point e.m.f.
Result of when a sinusoidal e.m.f. is applied to the secondary terminals of a current transformer is increased by 10% causes the exciting current to increase by 50%.

Knob and Tube
A system of wiring in which individual, loom-covered hot and neutral conductors were run using porcelain knobs to support the wires along framing members and porcelain tubes to protect wires passing through framing members.

Knockout (K.O.)
A partially prepunched opening in a box that is removed to allow the entry of cable. A knockout that is mistakenly opened or is open because a cable is removed must be filled with a knockout seal.

Labeled

Items to which a label, trademark, or other identifying mark of nationally recognized testing labs has been attached to identify the items as having been tested and meeting appropriate standards.

Lag

The condition where the current is delayed in time with respect to the voltage in an ac circuit (for example, an inductive load).

Lagging Load

An inductive load with current lagging voltage. Since inductors tend to resist changes in current, the current flow through an inductive circuit will lag behind the voltage. The number of electrical degrees between voltage and current is known as the "phase angle". The cosine of this angle is equal to the power factor (linear loads only).

Lambert

An old unit of luminance. The luminance of a uniform diffuser of light which emits one lumen per square centimetre. 1 lambert = 10^{-4} lux

Laminations

Thin layers or sheets. The term refers to the thin pieces of iron used to build up the core of a transformer.

Lamp

A complete light source unit, usually consisting of a light generating element (arc tube or filament), support hardware, enclosing envelope and base.

Lamp Lumen Depreciation, LLD (Lighting)

Information about the chosen lamp and its lumen depreciation and mortality are available from lamp manufacturers' literature. Rated average life should be determined for the specific hours per start; it should be known when burnouts occur in the lamp life cycle. From these facts, a practical, group relamping cycle should be established. Based on the hours elapsed to lamp replacement, the LLD factor can be determined.

Lampholder

A threaded adapter that converts the thread size of the lampholder in which it is inserted so that the lampholder can accept an incandescent lamp bulb of a different size thread.

Langley (L)

Unit of solar irradiance. One gram calorie per square centimeter. 1 L = 85.93 kwh/m2.

Laplace Transform

An integral transformation of a function from the time domain to the complex frequency domain. Used to analyse transients in circuits.

Laser

Acronym for Light Amplification by Stimulated Emission of Radiation. The laser produces a powerful, highly directional, monochromatic, and coherent beam of light.

Laser Driver

An IC that supplies modulated current to a laser diode in response to an input serial-data stream.

Latch

A non-clocked flip-flop.

Latent Heat

Quantity of heat required to effect a change of state of a unit mass of a substance from solid to liquid (latent heat of fusion) or from liquid to vapour (latent heat of vaporisation) without change of temperature.

Lateral Circuit

A tap-off line to take primary distribution from the main power line to a nearby load center.

Lateral Light Distribution

Lateral light distributions are classified by IES distribution Types I, II, III, IV and V. In general, the larger the number, the more is projected across the roadway. This allows the lighting designer to select the appropriate distribution pattern for a given roadway width.

Lattice

The regular geometrical arrangement of points in crystal space.

Lattice Parameter

The combination of unit cell edge lengths and interaxial angles that defines the unit cell geometry.

Lay Direction

1) The direction in which the wires of a conductor are twisted. 2) The twist of conductors in a cable.

Lay Length

The distance required to complete one revolution of helically laid strands of wires.

Layoff

Excess capacity of a generating unit, available for a limited time under the terms of a sales agreement.

LC Circuit

An electrical network containing both inductive and capacitive elements.

Leaching

Washing out of a soluble constituent of a material.

Lead

The condition where the current precedes in time with repect to the voltage in an ac circuit (for example, a capacitive load).

Lead Acid Battery

The assembly of one or more cells with an electrolyte based on dilute sulfuric acid and water, a positive electrode of lead dioxide and negative electrodes of lead. Lead Acid batteries all use the same basic chemistry. The positive plate is comprised of lead dioxide and the negative of finely divided lead. Both of these active materials react with the sulfuric acid electrolyte to form lead sulfate on discharge. The reaction is reversed on recharge. Batteries are constructed with lead grids to support the active material and individual cells are connected to produce a battery in a plastic or glass case.

Lead Dioxide (Battery)

The higher oxide of lead present in charged positive

plates. It is frequently referred to as lead peroxide.

Lead Sulfate
A lead salt formed by the action of sulfuric acid on lead oxide during paste mixing and formation. It is also formed electromechanically when a battery is discharged.

Leading Edge
The edge of a blade that faces toward the direction of rotation.

Leading Load
A capacitive load with current leading voltage. Since capacitors resist changes in voltage, the current flow in a capacitive circuit will lead the voltage.

Leadless Ceramic Chip Carrier
1. Leadless Ceramic Chip Carrier or Leadless Chip Carrier: An IC package, usually ceramic, that has no leads (pins). It instead uses metal pads at its outer edge to make contact with the printed circuit board.
2. Leaded Chip Carrier, also called PLCC or Plastic Leaded Chip Carrier: A square surface mount chip package in plastic with leads (pins) on all four sides.

Leakage Current
Electric current in an unwanted conductive path under normal operating conditions. Current flowing from the enclosure, or parts thereof, accessible to the operator in normal use, through an external conductive connection other than the protective earth conductor to earth or to another part of the enclosure.

Leakage Factor
Ratio of the total flux to the useful flux is defined as the leakage factor. It has a value of around 1.2 for electrical machines.

Leakage Inductance
An inductive component present in a transformer that results from the imperfect magnetic linking of one winding to another.

Leased Line
A dedicated circuit, typically supplied by the telephone company, that permanently connects two or more user locations. These lines are used to transmit data.

Least Cost Alternatives

The lowest cost option for providing for incremental demands. In least cost planning to serve electric demands, the least cost alternatives are often construed broadly to include demand-side management as well as various generation and purchased power options.

Least Significant Bit (LSB)

In a binary number, the LSB is the least weighted bit in the number. Typically, binary numbers are written with the MSB in the left-most position; the LSB is the furthest-right bit.

Light Emitting Diode (LED)

Light Emitting Diode, emits light when current is passed through it.

Leeward

Away from the direction from which the wind blows.

Left Hand Generator Rule

A method of determining the direction of current flow (polarity) in a conductor, as follows: Extend the thumb first finger, and second finger of the left hand at right angles to each other. When the thumb is pointed in the direction (upward or downward) of conductor motion as it passes through the magnetic field, and the first finger is pointed to the south pole of the magnetic field (matching the direction of magnetic flux), the second finger will point in the direction of current flow (toward the negative pole or terminal).

Lenz's Law

When a coil and a magnetic field move relative to each other, the electric current is induced in the coil will have a magnetic field opposing the motion.

Level Translator

A device which translates a logic signal from one type to another, for example, ECL to TTL.

Levelized

A lump sum that has been divided into equal amounts over period of time.

Lever Rule

Mathematical expression whereby the relative phase amounts in a two-phase alloy at equilibrium my be computed.

Leverage Ratio
A measure that indicates the financial ability to meet debt service requirements and increase the value of the investment to the stockholders (i.e., the ratio of total debt to total assets).

Liability
An amount payable in dollars or by future services to be rendered.

Life
The period during which a system is capable of operating above a specified performance level.

Life-Cycle Cost
The estimated cost of owning and operating a photovoltaic system for the period of its useful life.

Lift
The force exerted by moving air on asymmetrically-shaped wind generator blades at right angles to the direction of relative movement. Ideally, wind generator blades should produce high Lift and low Drag.

Light
Any radiation capable of causing a visual sensation direct i.e. Visible electromagnetic radiation in the wavelength range 400 to 700 nano meter.

Light Emitting Diode (LED)
Emits light when current is passed through it. A semiconductor light source that radiates light such as red, green, yellow and white, or invisible light such as infra red.

Light Loss Factor (LLF)
A factor used in calculating illuminance after a given period of time and under given conditions. It takes into account temperature and voltage variations, dirt accumulation on luminaire and lit surfaces, lamp depreciation, maintenance procedures and atmospheric conditions. It was formerly referred to as maintenance factor.

Light Oil
Lighter fuel oils distilled off during the refining process. Virtually all petroleum used in internal combustion and gas turbine engines is light oil.

Light Traffic
A grade level Box or Cover rating.

Light Trapping

The trapping of light inside a semiconductor material by refracting and reflecting the light at critical angles; trapped light will travel further in the material, greatly increasing the probability of absorption and hence of producing charge carriers.

Light Year

A measure of astronomical distances. 1 light year = 9.461 x 10^{15} m

Lighted (Illuminated)

A receptacle with a face that becomes illuminated when the device is connected to an energized electrical circuit.

Lighted Handle

A switch with an integral lamp in its actuator (toggle, rocker or pushbutton) that illuminates when the switch is connected to an energized circuit and the actuator is in the OFF position.

Lighting Arrester

A device that protects power lines and equipment against high voltage lighting surges and switching surges. Connected from line to ground potential, the device has a very high resistance to current flow at normal voltages but when a very high voltage surge hits it, it becomes a very low resistance, passing damaging surges and current to ground.

Lighting Maintenance Factor (MF)

The result of time-dependent depreciation effects must be considered in the initial design. Regular maintenance is particularly important with regard to energy conservation and these plans, once incorporated into the design, should be carried out or the system will not perform as expected.

Lightning

Any form of visible electric discharge between thunder clouds or between a thunder cloud and the earth. Ninety percent of all lightning never touches the ground—it occurs inside the thunder cloud or jumps from cloud to cloud.

Lightning & Switching Impulses

A distinction is made between Lightning and Switching impulses on the basis of duration of the wave front. Impulses with wave-front durations of up to a few tens of microseconds are in general considered to be lightning impulses. Those having durations of tens to thousands of microseconds are considered to be switching surges.

Lightning Arrestor

This protects lines, transformers, and equipment from lightning surges by carrying the charge to the ground. Lightning arrestors serve the same purpose on a line as a safety valve on a steam boiler.

Lightning Rod

A grounded metallic rod set up on a structure (like a building) to protect it from lightning.

Lignite

The lowest rank of coal, often referred to as brown coal, used almost exclusively as fuel for steam-electric power generation. It is brownish-black and has a high inherent moisture content, sometimes as high as 45 percent. The heat content of lignite ranges from 9 to 17 million Btu per ton on a moist, mineral-matter-free basis. The heat content of lignite consumed in the United States averages 13 million Btu per ton, on the as-received basis (i.e., containing both inherent moisture and mineral matter).

Limit Switch

A switch that is operated by some part or motion of a power-driven machine or equipment to alter the electric circuit associated with the machine or equipment.

Limited Approach Boundary

An approach limit at a distance from an exposed live part within which a shock hazard exists.

Limiting Value of the output current

The upper limit of the output current which cannot, by design be exceeded under any conditions.

Line

A designation of one or more power-carrying conductors for power distribution. The brown (or red) wire is the line conductor, the blue (or black) wire is the neutral, and the green-yellow (or green) wire is the ground. The voltage difference between the line conductor and the neutral is the supply voltage.

Line/Wire Loss

Refers to the voltage or power lost due to the resistance of any wire (or wires) in any electrical circuit.

Line Conditioner

This term isn't used consistently, therefore its meaning has been blurred. The term is sometimes used to describe equipment that provides some type of filtering or Regulation to an AC power source and may be any of the following devices: Surge Suppressor, Ferroresonant Transformer, AC Filter or Tap Changing Regulator.

Line Drop Compensator (LDC)

A Line Drop Compensator is utilized to provide constant voltage at the load.

Line Hose

A rubber dielectric cover for conductor that is used to electrically isolate a worker from an energized conductor. Line hose is made by W.H. Salisbury & Company.

Line Imbalance

Unequal loads on the phase lines of a multiphase feeder.

Line Interactive UPS

A type of uninterruptible power supply (UPS) that is in standby mode most of the time but contains a transformer that helps condition and adjust the voltage until the voltage exceeds a set limit. If the voltage limit is exceeded, the UPS can support the load until the voltage returns to within the limits or the battery is drained.

Line Loss

Electric energy lost in the process of transmitting it over power lines.

Line Reactors

An inductor, or coiled conductor, that opposes any changes in the current flow.

Line Regulation

The ability of a power-supply voltage regulator to maintain its output voltage despite variations in its input voltage.

Line to Line

A term used to describe a given condition between conductors of a multiphase feeder.

Line to Neutral

A term used to describe a given condition between a phase conductor and a neutral conductor.

Line Voltage

The voltage between two lines (or phases) of a three phase system is defined as the line-to-line voltage or more commonly as the line voltage. For a balanced three phase system, the line voltage is Ó3 times the phase-to-neutral voltage.

Linear

Having the property that the output is proportional to the input

Linear Circuit

One whose output is linearly related to the input. A circuit which obey's Ohm's law.

Linear Current Booster (LCB)

An electronic circuit that matches PV output directly to a motor. Most commonly used in direct water pumping.

Linear Feedback Shift Register (LFSR)

A shift register in which some of its outputs are connected to the input through some logic gates (typically, an exclusive-or (XOR). A wide variety of bit patterns can be generated inexpensively, including pseudo-random sequences. Can be used as a noise generator.

Linear Load

A load in which the current relationship to voltage is constant based on a relatively constant load impedance.

Linear Mode

Uses a linear-pass element (BJT or FET) to control/regulate the charging voltage/current.

Linear Motor

A form of motor (normally induction motor) in which the stator and rotor are linear instead of cylindrical and parallel instead of co-axial.

Linear Regulator

A voltage regulator that is placed between a supply and the load and provides a constant voltage by varying its effective resistance.

Line-Commutated Inverter

An inverter that is tied to a power grid or line. The commutation of power (conversion from direct current to alternating current) is controlled by the power line, so that, if there is a failure in the power grid, the photovoltaic system cannot feed power into the line.

Liner

Cloth gloves used to line the inside of a rubber insulating glove.

Liquid

A state of matter between a solid and a gap. Liquids assume the shape of the vessel containing them, other than the top surface which will assume a horizontal position when free to air. They are only slightly compressible.

Liquid Crystal Display (LCD)

Displays made from liquid crystals which usually change from transparent to opaque in the presence of electric or magnetic fields. They are commonly used in digital displays.

Liquid Electrolyte Battery

A battery containing a liquid solution of acid and water. Distilled water may be added to these batteries to replenish the electrolyte as necessary.

Also called a flooded battery because the plates are covered with the electrolyte.

Liquid-Filled Transformer

A transformer in which the core and coil are immersed in a liquid which acts as both a cooling and insulating medium.

Lissajous Pattern

The pattern appearing on an oscilloscope when harmonically related signals are applied to the horizontal and vertical inputs of an oscilloscope.

Listed

Equipment or materials included in a list published by an organization acceptable to the authority having jurisdiction and concerned with product evaluation, that maintains periodic inspection of production of listed equipment or materials, and whose listing states either that the equipment,or material meets appropriate designated standards or has been tested and found suitable for use in specified manner.

Lithium batteries

Lithium batteries for low-power, high-reliability, long-life applications such as non-volatile memory and timekeeping (typically in coin-shaped cells) use a variety of lithium-based chemistries (as differentiated from lithium-ion).

Dallas Semiconductor NV SRAM and timekeeping products use mostly BR chemistry (poly-carbonmonofluoride) primary (non-rechargeable) lithium coin cells. We use CR chemistry (manganese dioxide) primary lithium coin cells in microcontroller and touch products. Some new products use "manganese lithium" (ML) chemistry, which is chemically close to the CR, but is a secondary (rechargeable) lithium coin cell.

Live Part

A conductor or conductive part intended to be energised in normal use, including a neutral conductor but, by convention, not a PEN conductor.

Load

Any passive electrical device connected to a power source may be called by the general term of "load". It is the amount of electric power delivered or required at any specific point or points on a system. The requirement originates at the energy consuming equipment of the consumers. [Unit: kW or MW].

Load (Electric)

The amount of electric power delivered or required at any specific point or points on a system. The requirement originates at the energy-consuming equipment of the consumers.

Load Balancing

Switching the various loads on a multi-phase feeder to equalize the current in each line.

Load Building

Refers to programs that are aimed at increasing the usage of existing electric equipment or the addition of electric equipment. Examples include industrial technologies such as induction heating and melting, direct arc furnaces and infrared drying; cooking for commercial establishments; and heat pumps for residences. Load Building should include programs that promote electric fuel substitution. Load Building effects should be reported as a negative number, shown with a minus sign.

Load Center

Source for all power to the home. All circuits originate from the "Load Center" or "Service Panel." Circuit breakers are located within this panel.

Load Circuit

The wire, switches, fuses, etc. that connect the load to the power source.

Load Curve

A curve showing instantaneous demand (kVA or MVA) versus time. Curves are usually plotted for one day or one week. Integrating the load curve will provide the amount of energy consumed.

Load Diversity

The condition that exists when the peak demands of a variety of electric customers occur at different times. This is the objective of "load molding" strategies, ultimately curbing the total capacity requirements of a utility.

Load Duration Curve

A curve that displays load values on the horizontal axis in descending order of magnitude against percent of time (on the vertical axis) the load values are exceeded.

Load Factor

The ration of average demand to peak demand. It is a measure of efficiency that indicates whether a system's electric use over a period of time is reasonably stable or if it has extreme peaks and valleys. A high load factor usually results in a lower average price per kilowatt-hour than a low load factor.

Load Fault

A malfunction that causes the load to demand abnormally high amounts of current from the source.

Load Forecast

Estimate of electrical demand or energy consumption at some future time.

Load Line

Locus of instantaneous operating points used to find the exact operating values of voltage and current.

Load Management

Refers to programs other than Direct Load Control and Interruptible Load that limit or shift peak load from on-peak to off-peak time periods. It includes technologies that primarily shift all or part of a load from one time-of-day to another and secondarily may have an impact on energy consumption. Examples include space heating and water heating storage systems, cool storage systems, and load limiting devices in energy management systems. This category also includes programs that aggressively promote time-of-use (TOU) rates and other innovative rates such as real time pricing. These rates are intended to reduce consumer bills and shift hours of operation of equipment from on-peak to off-peak periods through the application of time differentiated rates.

Load Profile

An allocation of electricity usage to discrete time intervals over a period of time, based on individual customer data or averages for similar customers. A load profile may be used to esti-mate electric supply requirements and determine the cost of service to a customer. Generally speaking, customers with small demand requirements may participate in electric customer choice using a load profile rather than interval demand meter data."

Load Regulation

A term used to describe the effects of low forward transfer impedance. A power conditioner with "load regulation" may not have voltage regulation. Removing the power conditioner altogether will improve load regulation.

Load Resistance

The resistance presented by the load.

Load Shape

A curve on a chart showing power (kW) supplied (on the horizontal axis) plotted against time of occurrence (on the vertical axis), and illustrating the varying magnitude of the load during the period covered.

Load Shifting

A load shape objective that involves moving loads from peak periods to off-peak periods. If a utility does not expect to meet its demand during peak periods but has excess capacity in the off-peak periods, this strategy might be considered.

Load Switching
Transferring the load from one source to another.

Load Unbalance
Unequal loads on the phase lines of a multi- phase system.

Loadbreak
Refers to a group of rubber insulating products used to electrically connect apparatus with which load can be separated manually. Loadbreak products are manufactured by T&B Elastimold.

Loading Effect
Connection of a meter (an ammeter in series or a voltmeter in shunt) to a circuit to make a measurement alters the original circuit in that it draws some energy from the circuit. The error caused is known as the loading effect.

Local Control Mode
When set for a given control

point it means that the commands can be issued from this point.

Local Lighting
Lighting for a specific visual task, additional to and controlled separately from the general lighting.

Local Multipoint Distribution Service (LMDS)
A broadband radio service, located in the 28GHz and 31GHz bands, designed to provide two-way transmission of voice, high-speed data and video (wireless cable TV). In the U.S., FCC rules prohibit incumbent local exchange carriers and cable-TV companies from offering in-region LMDS.

Local Roadway (Lighting)
Roadways used primarily for direct access to residential, commercial, industrial or other abutting properties. They do not include roadways carrying through traffic. Long local roadways will generally be divided into short sections by collector roadway systems.

Local Temperature
The temperature measured on the die of the tempera-

ture-measuring integrated circuit.

Local Temperature Sensor
An element or function of an integrated circuit that measures its own die temperature.

Location, damp
A location subject to moderate amount of moisture such as some basements, barns, cold storage, warehouse and the like.

Location, dry
A location not normally subject to dampness or wetness: a location classified as dry may be temporarily subject to dampness or wetness, as in case of a building under construction.

Location, wet
A location subject to saturation with water or other liquids.

Locked-Rotor Current
Steady-state current taken from the line with the rotor of a motor at standstill and at rated voltage and frequency.

Locked-Rotor Torque
The minimum torque that a

motor will develop at standstill for all angular positions of the rotor, with rated voltage applied at rated frequency.

Lockout
A mechanical device which may be set to prevent the operation of a push-button or other device.

Locus
The line that can be drawn through adjacent positions satisfying a given criteria.

Logarithmic Pot
Potentiometer in which the resistive element is logarithmic rather than linear. Most commonly used for audio volume controls because the ear responds logarithmically.

Logic Circuit
A circuit that behaves according to a set of logic rules.

Logic Level
State of a voltage variable. States HIGH and LOW correspond to the two usable voltage levels of a digital device.

Long Distribution (Lighting)
A luminary is classified as

having a long light distribution when its max candlepower point falls between 3.75MH – 6.0MH TRL. The maximum luminaire spacing-to-mounting height ratio is generally 12.0 or less.

Long Haul

A network that spans distances larger than a local area network (LAN). Because electrical and optical transmissions fade over distance, long-haul networks are difficult and expensive to implement.

Long-Term Stability

The stability over a period of one year.

Loop Current

A component of current common to the complete loop.

Loop or Mesh

A closed path of elements in a circuit.

Loss

Power expended without accomplishing useful work.

Loss Angle

Angle of deviation of the current from the ideal current for a dielectric for a sinusoidal input.

Loss Factor

A factor which defines the loss component of a dielectric.

Loss of Load Probability (LOLP)

A measure of the probability that system demand will exceed capacity during a given period; this period is often expressed as the expected number of days per year over a long period, frequently taken as ten consecutive years. An example of LOLP is one day in ten years.

Loss Tangent

The loss factor, which also corresponds to the tangent of the loss angle.

Losses

The general term applied to energy (kWh) and capacity (kW) lost in the operation of an electric system. Losses occur principally as energy transformations from kWh to waste-heat in electrical conductors and apparatus. This waste-heat in electrical conductors and apparatus. This power expended without accomplishing useful work occurs primarily on the transmission and distribution system.

Low Drop Out (LDO)

A linear voltage regulator that will operate even when the input voltage barely exceeds the desired output voltage.

Low Emission Vehicle

All new cars sold in California starting in 2004 will have at least a LEV or better emissions rating.

Low Heat Value (LHV)

The low or net heat of combustion for a fuel assumes that all products of combustion, including water vapor, are in a gaseous state.

Low Noise Earth

An earth connection in which the level of conducted or induced interference from external sources does not produce an unacceptable incidence of malfunction in the data processing or similar equipment to which it is connected. The susceptibility in terms of amplitude/frequency characteristics varies depending on the type of equipment.

Low Pass Filter

A filter designed to pass only frequencies from d.c. up to the cut-off frequency.

Low Speed Vehicle (LSV)

Another name for Neighborhood Electric Vehicles rating.

Low Voltage

A wiring system that provides power to some electronic devices operating on a voltage level much lower than the standard 110 volts. Such devices might be doorbells and thermostats.

Low Voltage Cutoff (LVC)

The voltage level at which a charge controller will disconnect the load from the battery.

Low Voltage Disconnect

The voltage at which a charge controller will disconnect the load from the batteries to prevent over-discharging.

Low Voltage Disconnect Hysteresis

The voltage difference between the low voltage disconnect set point and the voltage at which the load will be reconnected.

Low Voltage Warning

A warning buzzer or light that indicates the low battery voltage set point has been reached.

Low-Pressure Mercury (Vapor) Lamp

Mercury vapor lamp, with or without a coating of phosphor, in which during operation the partial pressure of the vapor does not exceed 100 pa.

Low-Side

An element connected between the load and ground. Low-side current sensing applications measure current by looking at the voltage drop across a resistor placed between the load and ground.

Low-Voltage

A switch rated for use on low-voltage circuits of 50 volts or less.

L-Rated

A switch specially designated with the letter "L" in its rating that is rated for controlling tungsten filament lamps on AC circuits only.

Lug

Used to terminate a wire.

Lumen (lm)

SI unit for measuring the flux of light. One lumen is equal to the luminous flux emitted in unit solid angle (steradian) by uniform point source having a luminous intensity of 1 candela.

Lumens Per Watt (LPW)

The ratio of light energy output (Lumens) to electrical energy input (Watts).

Luminaire

A complete lighting unit consisting of a light source with a means of distribution (reflector and/or refractor), lamp positioning (socket), lamp protection (housing) and a provision for power connection.

Luminaire Dirt Depreciation (LDD)

The accumulation of dirt on luminaires results in a loss of light output on the road. This loss is known as the LDD factor and is determined by estimating the dirt category from the graph below. From the appropriate dirt condition curve and the proper elapsed time of the planned cleaning cycle, the LDD factor is then found.

Luminaire Efficiency

The ratio of total lumen output of a luminaire to the lumen output of the lamps, expressed as a percentage.

Luminance

In a direction and at a point of a real or imaginary surface

The quotient of the luminous flux at an element of the surface surrounding the point, and propagated in directions defined by an elementary cone containing the given direction, multiplied by the solid angle of the cone and the area of the orthogonal projection of the element of the surface on a plance perpendicular to the given direction.

Luminous Efficacy

Quotient of the luminous flux emitted by a source and the power consumed. [Unit lumen per watt, lm/W] Visible light output of a lu-minary measured relative to power input.

Luminous Flux

The quantity derived from radiant flux by evaluating the radiation according to its action upon the standard photometric observer. [Unit lumen, lm].

Luminous Intensity Distribution

Distribution of the luminous intensities of a lamp or luminaire in all spatial directions.

Lux (lx)

SI unit for measuring the illumination of a surface. One lux is defined as an illumination of one lumen per square meter.

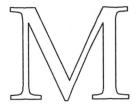

Machine Tool Wire (MTW)

Machine tool wire, used for electrical connections inside equipment.

Macromolecule

A huge molecule made up of thousands of atoms.

Made Circuit

A closed or completed circuit.

Magnetic Field

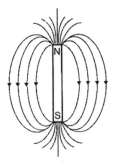

A region of space that surrounds a moving electrical charge or a magnetic pole, in which the electrical charge or magnetic pole experiences a force that is above the electrostatic ones associated with particles at rest.

Magnetic Field Strength

The intensity of an externally applied magnetic field.

Magnetic Flux

A measure of quantity of magnetism, taking account of the strength and the extent of a magnetic field. [Unit: weber]

Magnetic Flux Density

The magnetic field produced in a substance by an external magnetic field.

Magnetic Induction

The process of setting up magnetism in an object that is placed in a magnetic field.

Magnetic Linens
The imaginary lines of force that make up a magnetic field.

Magnetic Mechanism
The magnetic mechanism uses a solenoid with an iron piece to operate the circuit breaker in the event of an overcurrent.

Magnetic Pole
The point at which the magnetic lines are concentrated. In every magnet, there is one north pole and one south pole.

Magnetic Shield
A piece of magnetic material used to carry the magnetic lines around and object to prevent the object from being affected by the magnetic field.

Magnetic Susceptibility
The proportionality constant between the magnetization M and the magnetic field strength H.

Magnetisation Curve
The relationship between the magnetic flux density and the applied magnetic field (or the magnetic flux and the applied mmf) is called the magnetisation curve.

Magnetism
The property of certain materials to attract iron and other magnetic materials.

Magnetite
A certain type of ore which, in its natural state, has the property of magnetism.

Magnetization
The total magnetic moment per unit volume of material. Also, a measure of the contribution to the magnetic flux by some material within an h field.

Magneto-Hydrodynamics (MHD)
A method of generating electricity by subjecting the free electrons in a high velocity flame or plasma to a strong magnetic field. The free electron concentration in the flame is increased by the thermal ionisation of added substances of low ionisation potential.

Magnetomotive Force
The force that sets up a magnetic field within and around an object.

Magnetostriction

A change in the dimensions of ferromagnetic substances on magnetisation.

Main Circuit Breaker

A circuit breaker that can interrupt all power in a facility.

Main Earthing Terminal

The terminal or bar provided for the connection of protective conductors, including equipotential bonding conductors, and conductors for functional earthing, if any, to the means of earthing.

Main Protection

The protection system which is normally expected to operate in response to a fault in the protected zone.

Main Switch

The principal (or main) switch in an electrical installation.

Maintained Contact

A switch where the actuator (toggle, rocker, pushbutton or key mechanism) makes and retains circuit contact when moved to the ON position. The contacts will only be opened when the actuator is manually moved to the OFF position. Ordinary light switches are maintained contact switches.

Maintenance Factor

Ratio of the average illuminance on the working plane after a specified period of use of a lighting installation to the average illuminance obtained under the same conditions for a new installation.

Maintenance-Free Battery

A sealed battery to which water cannot be added to maintain electrolyte level.

Major Roadway (Lighting)

That part of the roadway system that serves the principal network for through-traffic flow. The routes connect areas of principal traffic generation and important rural highways entering the city.

Majority Carrier

Current carriers (either free electrons or holes) that are in excess in a specific layer of a semiconductor material (electrons in the n-layer, holes in the p-layer) of a cell.

Make-Before-Break (MBB)

In a switching device, a configuration in which the new connection path is established before the previous contacts are opened. This prevents the switched path from ever seeing an open circuit.

Applies to mechanical systems (e.g. that use relays or manual switches) and to solid-state analog multiplexers and switches.

Malleability

Capacity of being hammered out into thin sheets.

Manchester Data Encoding

Manchester encoding is a form of binary phase-shift keying (BPSK) that has gained wide acceptance as a modulation scheme for low-cost radio-frequency (RF) transmission of digital data. Its key characteristic is that it encodes data in a way that insures there will never be long strings of continuous zeros or ones. The guaranteed transitions means that the clock can be derived from the transmitted data, allowing the link to function with variable signal strengths from transmitters with imprecise, low-cost, data-rate clocks.

Manual Motor Controller

A switch designed for controlling small DC or AC motor loads, without overload protection.

Manual Reset

A starter that automatically deactivates a failed fluorescent lamp to eliminate flickering. A reset button provides a means of activating the circuit after lamp replacement.

Manual Transfer Switch

A switch designed so that it will disconnect the load from one power source and reconnect it to another source while at no time allowing both sources to be connected to the load simultaneously.

Marginal Cost

The sum that has to be paid for the next increment of product of service. The marginal cost of electricity is the price to be paid for kilowatt-hour above and beyond those supplied by presently available generating capacity irrespective of sunk costs.

Market Clearing Price

The price at which supply equals demand for the Day Ahead and/or Hour Ahead Markets.

Market Eligibility

The percentage of equipment still available for retrofit to the demand-side management measure. For example, if 20 percent of customers where demand controllers are feasible have already purchased demand controllers, then the eligible market eligibility factor is 80 percent.

Market Power

The ability of a seller/buyer, either individually or in collaboration with other sellers/buyers, to affect the price of electricity in the market.

Marketer

An agent for generation projects who markets power on behalf of the generator. The marketer may also arrange transmission, firming or other ancillary services as needed. Though a marketer may perform many of the same functions as a broker, the difference is that a marketer represents the generator while a broker acts as a middleman.

Martensite

A metastable Fe-C composition consisting of supersaturated carbon in iron that is the product of a diffusionless (athermal) transformation from austenite.

Matching Transformer

A device used to convert impedance between two levels. A common use is between a 75 ohm impedance and a 300 ohm impedance.

Matrix

The body constituent of a composite or two-phase alloy that completely surrounds the dispersed phase and gives the body its bulk form.

Max. Hold Step (MV)

When switching between sample mode and hold mode, charge injection from stray capacitance causes the maximum voltage of the hold capacitor to change.

Maximum (Peak) Surge Current

The peak surge current a Surge Protective Device can withstand, based on IEEE Standard C62.45 test waveforms.

Maximum Demand

The largest of all demands of the load (usually expressed in kVA or MVA) that has occurred within a specified period of time.

Maximum Operating Voltage

This is the maximum 50 to 60 Hz AC voltage the unit can sustain without damage or failure of the suppressor.

Maximum Permissible Values of the input current and voltage

Values of current and voltage assigned by the manufacturer which the transducer will withstand indefinitely without damage.

Maximum power

Also referred to as peak power. The point on a device's I-V curve where the product of I and V (Pmax, measured in watts) is maximized. The points on the I and V scales which describe this curve point are named Imp (current @ max power) and Vmp (voltage @ max power.).

Maximum Power Point (MPP)

The point on the current-voltage (I-V) curve of a module under illumination, where the product of current and voltage is maximum. [UL 1703] For a typical silicon cell, this is at about 0.45 V.

Maximum Power Point Tracker (MPPT)

A power conditioning unit that automatically operates the PV-generator at its MPP under all conditions. An MPPT will typically increase power delivered to the system by 10% to 40%, depending on climate conditions and battery state of charge.

Maximum Power Tracking

Operating a photovoltaic array at the peak power point of the array's I-V curve where maximum power is obtained. Also called peak power tracking.

Maxwell (Mx)

An old unit of magnetic flux. $1 \text{ Mx} = 10^{-8} \text{ Wb}$

Mean Time Between Failure

A statistical estimate of the time a component, subassembly, or operating unit will operate before failure will occur.

Mean-Sensing Transducer

A transducer which measures the mean or average value of the input waveform but which is adjusted to give an output corresponding to the r.m.s. value of the input when that input is sinusoidal.

Measurand
A quantity subjected to measurement.

Measure Life
The length of time that the demand-side management technology will last before requiring replacement. The measure life equals the technology life. These terms are used synonymously.

Measured Limiting (used to be known as "let-through") Voltage
This is the maximum voltage measured across the terminals of the suppressor during the time the testing voltages were applied to the unit..

Measurement Systems Analysis (MSA)
Measurement Systems Analysis is a method for ensuring product test measurements are reliable, robust, and of good statistical merit.

Measuring Element
A unit or module of a transducer which converts the measurand, or part of the measurand into a corresponding signal.

Measuring Range
The part of the span where the performance complies with the accuracy requirements.

Measuring Relay
An electrical relay intended to switch when its characteristics quantity, under specified conditions and with a specified accuracy attains its operating value.

Mebi (Mi)
Binary multiple prefix corresponding to megabinary or 220 or (210)2 or 10242. [IEC 1998]

Media Access Control (MAC) Address
A hardware address that uniquely identifies each node of a network, as in IEEE-802 (Ethernet) networks. The MAC layer interfaces directly with the network medium.

Medical Electrical Equipment
Electrical equipment, pro-

vided with no more than one connection to a particular supply mains, and intended to diagnose, treat or monitor the patient under medical supervision and which makes physical or electrical contact with the patient and/or transfers energy to or from the patient and/or detects such energy transfer to or from the patient. The equipment includes those accessories as defined by the manufacturer which are necessary to enable the normal use of the equipment.

Medium Base

Same as the Edison base lampholder. An internally-threaded lampholder, with the inner shell approx. 1" in diameter. Designed for widely-used standard medium base lamps.

Medium Bi-Pin

A fluorescent lampholder with two contacts, used in pairs. For type T-8 tubular fluorescent lamps, approx. 1" in diameter.

Medium Distribution (Lighting)

A luminary is classified as having a medium light dis-tribution when its max candlepower point falls between 2.25MH 3.75MH TRL. The maximum luminaire spacing-to-mounting height ratio is generally 7.5 or less.

Medium Voltage

An electrical system or cable designed to operate between 1kv and 38kv.

Mega (M)

Decimal multiple prefix corresponding to a million or 10^6.

Mega Ohm

A unit of electrical resistance equal to one million ohms.

Megabits Per Second (Mbps)

A megabit is roughly a million bits of data. This abbreviation is used to describe data transmission speeds, such as the rate at which information travels over the internet.

Megahertz (MHz)

A frequency of one-million Hertz, or 10^6 cycles per second.

Mega-Volt

A unit of pressure equal to one million volts.

Megawatt (MW)

One thousand kilowatts (1,000 kW) or one million watts (1,000,000 watts). A term that is most often used to measure the capacity of a power plant.

Megawatt-hour (MWH)

Equal to 1,000 kilowatt-hours or 1 million watt-hours.

Megger

A test instrument for measuring the insulation resistance of conductors and other electrical equipment; specifically, a mega ohm (million ohms) meter; this is a registered trade mark of the James Biddle Co.

Megohmmeter

A testing device that applies a DC voltage and measures the resistance (in millions of ohms) offered by conductor's or equipment insulation.

Melt Time

The time needed for a fuse element to melt, thereby initiating operation of the fuse. Also known as Fuse Melt Time.

Melting Point

The constant temperature at which the solid and liquid phase of a substance are in equilibrium at a given pressure.

Member System

An Eligible Customer operating as a part of a lawful combination, partnership, association, or joint action agency composed exclusively of Eligible Customers.

Memory Effect

A phenomenon in which a cell, operated in successive cycles to less than full, depth of discharge, temporarily loses the remainder of its capacity at normal voltage levels (usually applies only to Ni-Cd cells). Note, memory effect can be induced in NiCd cells even if the level of discharge is not the same during each cycle. Memory effect is reversible.

Mer

The group of atoms that constitutes a polymer chain repeat unit.

Merchant Plant

An electric generator not owned and operated by an electric utility and that sells its output to wholesale and/ or retail customers. Merchant plants may also be called non-utility generators, or independent power producers. Merchant plants that sell all of their output wholesale must receive authorization from FERC to sell their output at market-based rates. As defined in Act 141 of 2000, Section 10g(d), a merchant plant means electric generating equipment and associated facilities with a capacity of more than 100 kW located in this state that are not owned and operated by an electric utility. Act 141, Section 10e(2), indicates that a merchant plant making sales to retail customers is an AES and must be licensed as such.

Mercury

A type of switch that uses mercury as the contact means for making and breaking an electrical circuit.

Mercury Lamps

An electric discharge lamp

in which the major portion of the radiation is produced by the excitation of mercury atoms.

Mercury Vapor Lamp (MV)

An HID light source in which the arc tube's primary internal element is Mercury Vapor.

Mesh Analysis

A method of analysis of circuits based on defining mesh currents as the variables.

Messenger

A bare wire used to support power or communications cables suspended overhead.

Metal

The electroposite elements and alloys based on these elements.

Metal Clad (Switchgear)

An expression used by some manufactures to describe a category of medium voltage switchgear equipment where the circuit breakers are all enclosed in grounded, sheet-steel enclosures. Such enclosures may be suitable for indoor use or may be enclosed in an integral weatherproof housing for installation out of doors.

Metal Enclosed

Surrounded by a metal case of housing, usually grounded.

Metal Enclosed (Switchgear)

An expression used by some manufacturers to describe a category of low voltage, 600 volt class switchgear equipment, where the circuit breakers are all enclosed in grounded, sheet-steel enclosures. Such enclosures normally are suitable only for indoor use.

Metal Halide Lamp

Discharge lamp in which the major portion of the light is produced by the radiation from a mixture of a metallic vapor (for example, mercury) and the products of the dissociation of halides (for example, halides of thallium, indium or sodium).

Metal Oxide Varistor (MOV)

A solid state device which becomes conductive when the voltage across it exceeds a certain level. When the voltage exceeds the MOV's threshold, a heavy current flows through the MOV instead of the load.

Metal Vapor Lamp

Discharge lamp such as the 'mercury (vapor) lamp' and the 'sodium (vapor) lamp' in which the light is mainly produced in a metallic vapor.

Metalclad

Devices in which the conducting parts are entirely enclosed in a metal casing.

Metastable

Nonequilibrium state that may persist for a very long time.

Meter (m)

The meter is the SI unit of length. It is a fundamental unit. It is defined as the length of the path travelled by light in vacuum during a time interval of 1/299 792 458 of a second [1983]

Meter Constant

This represents the ratio between instrument transformers (CTs, PTs) and the meter. It is used as a multiplier of the difference between meter readings to determine the kWh used. The meter constant is also used as a multiplier of the demand reading to determine the actual demand.

Meter, Kilogram and the Second (MKS system)

The metric system in which the meter, kilogram and the second are the fundamental units.

Metering

Monitoring of energy or water consumption or other data over a period of time.

Metering (non-tariff)

Values computed depending on the values of digital or analog inputs during variable periods.

Metering (tariff)

Energy values computed from digital and/or analog inputs during variable periods and dedicated to energy measurement for billing purposes.

Micro (μ)

A metric prefix meaning one millionth of a unit or 10^{-6}.

Micro Electronic Mechanical Systems (MEMS)

Systems that combine mechanical and electrical components and are fabricated using semiconductor fabrication techniques. Common examples are pressure and acceleration sensors which combine the sensor and amplification or conditioning circuitry. Other applications include switches, valves, and waveguides.

Micro Hydro

Hydro power systems with a power output of less than 100kW.

Microampere, or microamp (uA)

A millionth of an Ampere. Ampere is the basic unit for measuring electrical current. Often written as uA, but the u is a plain-text substitute for the Greek letter mu.

Microgroove

A small groove scribed into the surface of a cell which is filled with metal for contacts.

Micrometer

One millionth of a meter (10^{-6} m).

Micron

A metric term meaning one millionth of a meter.

Microphone (condenser)

An electromechanical transducer that converts sound pressure into an electrical signal. Condenser mics require power to charge a capacitor used to sense changes in sound pressure levels (spl).

Microprocessor

A large scale integrated circuit that can be programmed to perform arithmetic and logic functions and to manipulate data.

Microprocessor Supervisor

A device that monitors a host microprocessor or microcontroller's supply voltage and, in some cases, its activity. It monitors for a fault condition and takes appropriate action, usually issuing a reset to the microprocessor.

Microstructure

The structural features of an alloy that are subject to observation under a microscope.

Microwave

Electromagnetic radiation with wavelengths ranging from very short radio waves to almost infra-red region. Wavelengths from 300 mm to 1 mm.

Mid Point Sectioning Substation

A substation located at the

electrical interface of two sections of electrified railway. It contains provision for the coupling of the sections electrically in the event of loss of supply to one section.

Midget

A connector designed with a smaller body diameter than standard connectors with a similar rating.

mil

A unit of small length in the imperial system equal to one-thousandth of an inch. $1 \text{ mil} = 2.54 \times 10^{-5}$ m

mile

Unit of distance in the imperial system. 1 mile = 1.609 km

Miles Per Hour (mph)

Unit of speed in the imperial system. 1 mph = 0.44704 m/s

Miller Indices

A set of three integers that designate crystallographic planes, as determined from reciprocals of fractional axial intercepts.

Miller-Bravis Indices

A set of four integers that designate crystallographic planes in hexagonal crystals.

milli (m)

Decimal sub-multiple prefix corresponding to one-thousandth or 10^{-3}.

Milliampere, or milliamp (mA)

1/1000 of an Ampere. Ampere is the basic unit for measuring electrical current.

Min Stable Closed Loop Gain

The minimum closed-loop gain for which the amplifer is stable.

Mineral

Substance occurring naturally in the Earth.

Mineral Oil

General name given to the various mixtures of natural hydrocarbons.

Miniature

Designed for the smallest available incandescent lamps with a screw-in base, approx. 3/8" dia. Widely used in flashlights and toys, etc.

Miniature Bi-Pin

Similar to medium bi-pin lampholders, but designed for type T-5 tubular fluorescent lamps, approx. 5/8" in diameter.

Miniature Circuit Breaker (MCB)

A device designed to perform the same function as a fuse but resettable. When the circuit breaker activates, or trips out, it disconnects the circuit. The fault condition must be found and rectified before the mcb can be reset. Usually for ratings less than 63 A.

Minimum Charge

The total of all the payments a customer will owe for electric services no matter how much electricity the customer uses during a billing period. Typically, it is the total of all customer service charges, demand charges, and any other fees that are assessed regardless of energy used during any one billing period.

Minority Carrier

A current carrier, either an electron or a hole, that is in the minority in a specific layer of a semiconductor material; the diffusion of minority carriers under the action of the cell junction voltage is the current in a photovoltaic device.

Minority Carrier Lifetime

The average time a minority carrier exists before recombination.

Minority Carriers

The type of current carriers, free electrons or holes, of which a given semiconductor contains the least.

Mitigation Equipment

Equipment, such as surge suppressors and uninterruptible power supplies (UPS), which prevents unusual electrical disturbances from affecting equipment.

Mixed Dislocation

A dislocation that has both edge and screw components.

Mobile Substation

This is a movable substation which is used when a substation is not working or additional power is needed.

Mobile Transformer

A transformer that often is mounted on a leak proof base and can be installed and operated in a semitrailer, box truck or sea freight container.

Moderator

A substance used in nuclear reactors to reduce the speed of fast neutrons produced by nuclear fission.

Modified Sine Wave

A waveform that has at least three states (i.e., positive, off, and negative). Has less harmonic content than a square wave.

Modular

Individual-section wallplates with different openings that can be configured into a multi-gang plate.

Modularity

The use of multiple inverters connected in parallel to service different loads.

Modulator

The process of varying some characteristic of one wave (carrier wave) in accordance with some characteristic of another wave.

Module

A number of PV cells connected together, sealed with an encapsulant, and having a standard size and output power; the smallest building block of the power generating part of a PV array. Also called panel.

Module Derate Factor

A factor that lowers the photovoltaic module current to account for field operating conditions such as dirt accumulation on the module.

Modulus of Elasticity

The ratio of stress to strain for a material under perfectly elastic deformation.

Mogul

The largest screw-in type lampholder, designed for mogul incandescent lamps with a screw base of approx. 11/2" dia. Used in street lights and numerous commercial/industrial applications.

Molded On

A connector that is factory molded to a length of flexible cord.

Mole (mol)

The mole is the SI unit of the amount of substance. It is defined as the amount of substance of a system which contains as many elementary entities as there are atoms in 0.012 kilogram of carbon 12; [1971]

Molecules

Although all matter is made from atoms, atoms should not be confused with molecules, which are the smallest part of a compound. Water, for example, is a compound, not an element. The smallest particle of water a molecule made of two atoms of hydrogen and one atom of oxygen. If the molecule of water is broken apart, it becomes two hydrogen atoms and one oxygen atom, and is no longer water.

Momentary Contact

A switch that makes circuit contact only as long as the actuator (toggle, rocker, pushbutton or key mechanism) is held in the ON position, after which it returns automatically to the OFF position. This is a "Normally Open" switch. A "Normally Closed" switch will break circuit contact as long as it is held in the OFF position, and then automatically return to the ON position. Available in "Center OFF" versions with both Momentary ON and Momentary OFF positions.

Momentary Overvoltage or Swell

An increase in voltage outside the normal tolerance for a few seconds or less. Voltage swells are often caused by sudden load decreases or turn-off of heavy equipment.

Momentary Rating
The rating of a device to withstand momentary, very high current, without incurring damge.

Monitoring and Evaluation Cost
Expenditures associated with the planning, collection, and analysis of data used to assess program operation and effects. It includes activities such as load metering, customer surveys, new technology testing, and program evaluations that are intended to establish or improve the ability to monitor and evaluate the impacts of DSM programs, collectively or individually.

Monochromatic Radiation
Radiation characterized by a single frequency. In practice, radiation of a very small range of frequencies that can be described by stating a single frequency.

Monolithic
Fabricated as a single structure.

Monomer
A molecule consisting of a single mer.

Monopoly
One seller of electricity with control over market sales.

Monopsony
The only buyer with control over market purchases.

Monotonic
A sequence increases monotonically if for every n, Pn + 1 is greater than or equal to Pn. Similarly, a sequence decreases monotonically if for every n, Pn + 1 is less than or equal to Pn.
In plain language, the value rises and never falls; or it falls and never rises.

Mortality Curve
A graphic representation of lamp burnout as a function of time.

Mortality Rate
The number of operating hours elapsed before a certain percentage of the lamps fail.

Metal-Oxide-Silicon Field Effect Transistor (MOSFET)
Metal oxide semiconductor FET often used in switching amplifier applications. This transistor provides extremely low power dissipation even with high currents.

Most-Significant Bit (MSB)
In a binary number, the MSB is the most weighted bit in the number. Typically, binary numbers are written with the MSB in the leftmost position; the LSB is the furthest-right bit.

Motion Resistant Conductor
ACSR with Motion Resistant Variable Profile.

Motor
A rotating machine that converts electrical power (either alternating current or direct current) into mechanical power.

Motor Generator Set
A motor generator set consists of an ac motor coupled to a generator. The utility power energizes the motor to drive the generator, which powers the critical load. Motor generator sets provide protection against noise and spikes, and, if equipped with a heavy flywheel, they may also protect against sags and swells.

Motor, DC, Compound-Wound
A combination of the shunt wound and series wound type, which combines the characteristics of both. Varying the combination of the two windings may vary characteristics. These motors are generally used where severe starting conditions are met and constant speed is required at the same time.

Motor, DC, Series-Wound
This type of motor speed varies automatically with the load, increasing as the load decreases. Use of series motor is generally limited to case where a heavy power demand is necessary to bring the machine up to speed, as in the case of certain elevator and hoist installations, for steelcars, etc. Series-wound motors should never be used where the motor can be started without load, since they will race to a dangerous degree.

Motor, Shunt-Wound
This type of motor runs practically constant speed, regardless of the load. It is the type generally used in commercial practice and is usually recommended where starting conditions are not usually severe. Speed of the shunt-wound motors may be regulated in

two ways: first, by inserting resistance in series with the armature, thus decreasing speed: and second, by inserting resistance in the field circuit, the speed will vary with each change in load: in the latter, the speeds is practically constant for any setting of the controller. This latter is the most generally used for adjustable-speed service, as in the case of machine tools.

Motor, Single-Phase Induction

This motor is used mostly in small sizes, where polyphase current is not available. Characteristics are not as good as the polyphase motor and for size larger that 10 HP, the line disturbance is likely to be objectionable. These motors are commonly used for light starting and for running loads up to 1/3 HP Capacitor and repulsion types provide greater torque and are built in sizes up to 10 HP.

Motor, Squirrel-Cage-Induction

The most simple and reliable of all electric motors. Essentially a constant speed machine, which is adaptable for users under all but the most severe starting conditions. Requires little attention as there is no commutator or slip rings, yet operates with good efficiency.

Motor, Synchronous

Run at constant speed fixed by frequency of the system. Require direct current for excitation and have low starting torque. For large motor-generators sets, frequency changes, air compressors and similar apparatus which permits starting under a light load, for which they are generally used. These motors are used with considerable advantage, particularly on large power systems, because of their inherent ability to improve the power factor of the system.

Motor, Wound-Rotor (Slip Ring) Induction

Used for constant speed-service requiring a heavier starting torque than is obtainable with squirrel cage type. Because of its lower starting current, this type is frequently used instead of the squirrel-cage type in larger sizes. These motors are also used for varying-speed-service. Speed varies with this load, so that they should not be used where constant speed at each adjustment is required, as for machine tools.

Motors

Electronic device used to move, switch, or adjust one or more of the systems within a dwelling.

Moulded Case Circuit Breaker (Mccb)

A circuit breaker with a moulded case originally designed for currents exceeding 100 A.

Moving Coil Meter

In this instrument, a moving coil is suspended between the poles of a permanent magnet. When a current is passed through the coil, the coil becomes an electromagnet and tries to align with the permanent magnet. The deflecting becomes proportional to the current.

Moving Iron Meter

An instrument working on the principle of a moving iron placed within an electromagnet getting an induced emf and with the deflection being proportional to the square of the current. The meter is calibrated with the square root of the deflection and hence has basically a non-linear scale.

Movistor

Metal Oxide Varistor. Used to protect electronic circuits from surge currents such as those produced by lightning.

Multi-Chip Module (MCM)

An integrated circuit package that contains two or more interconnected chips.

Multicrystalline

Material that is solidified at such as rate that many small crystals (crystallites) form. The atoms within a single crystallite are symmetrically arranged, whereas crystallites are jumbled together. These numerous grain boundaries reduce the device efficiency. A material composed of variously oriented, small individual crystals. (Sometimes referred to as polycrystalline or semicrystalline).

Multi-Element Transducer

A transducer having two or more measuring elements.

The signals from the individual elements are combined to produce an output signal corresponding to the measurand.

Multi-Gang

A wallplate that has two or more gangs.

Multijunction Device

A photovoltaic device containing two or more cell junctions, each of which is optimized for a particular part of the solar spectrum, to achieve greater overall efficiency.

Multipath

In radio transmission, multipath refers to the simultaneous reception of two copies of the signal, that arrive via separate paths with different delays.

A common example is when a signal bounces off a building or other object and is received along with the direct (unbounced) signal. In television reception, this causes "ghosting" — one sees a faded echo on the screen horizontally displaced from the main image.

Another common example is in radio (especially AM radio), where the signal bounces off the ionosphere and one receives that delayed signal along with the directly transmitted signal. Usually, multipath is an undesired effect but in MIMO systems, separate antennas deliberately send replicas and sophisticated receivers piece together the fragments to improve performance.

Multiple Input, Multiple Output (MIMO)

A system has multiple antennas and multiple radios. It takes advantage of multipath effects, where a transmitted signal arrives at the receiver through a number of different paths. Each path can have a different time delay, and the result is that multiple instances of a single transmitted symbol arrive at the receiver at different times.

Usually multipath is a source of interference, but MIMO systems use the fact that data will arrive at the receiver at different times through different paths to improve the quality of the data link. For example, rather than relying on a single antenna path to receive an entire message, the message can be pieced together based on fragments received at the various antennas. This can act to either increase the data rate at a given range, or increase system range for a given data rate.

MIMO is used in the implementation of the 802.11n standard.

Multiplex

To put information from several sources on to a single line or transmission path.

Multi-Section Transducer

A transducer having two or more independent measuring circuits for one or more functions.

Multi-Shot Reclosing

A re-closing scheme that permits more than one re-closing operation of a CB after a fault occurs before lock-out occurs.

Multi-Stage Controller

A charging controller unit that allows different charging currents as the battery nears full state_of_charge.

Municipal Electric Utility

A power utility system owned and operated by a local jurisdiction or local authority.

Mutual Inductance

The ability of one conductor to induce an emf in a nearby conductor when the current in the first conductor changes. It is the constant of proportionality between the induced voltage in the second inductor and the rate of change of current in the first inductor.

Mutual Induction

The electromagnetic induction produced by one conductor in another nearby conductor, due to the moving flux of the first circuit cutting the conductors of the second circuit.

Nacelle

The protective covering over a generator or motor.

Nameplate Rating

The normal maximum operating rating applied to a piece of electrical equipment. This can include Volts, Amps, horsepower, kW, or any other specific item specification for the equipment.

NAND Gate

A	B	Output
0	0	1
0	1	1
1	0	1
1	1	0

Equivalent gate circuit

The logic gate that outputs a 0 only when all its inputs are 1s. It gives the complement of the AND gate.

Nano (n)

A metric prefix meaning one billionth of a unit or 10^{-9}.

Nanovolt (nV)

Unit of measure. A billionth of a volt.

Native Load Customers

The wholesale and retail customers on whose behalf the Transmission Provider, by statute, franchise, regulatory requirements, or contract, has undertaken an obligation to construct and operate the Transmission Provider's system to meet the reliable electric needs of such customers.

Natural Gas

A naturally occurring mixture of hydrocarbon and nonhydrocarbon gases found in porous geological formations beneath the earth's surface, often in association with petroleum.

The principal constituent is methane.

Natural Magnet

A material which, in its natural state, possesses the qualities of a magnet. Magnetite is a natural magnet.

Natural Monopoly

When the cost of utility service, such as gas, water or electric service, is minimized to customers if a single enterprise is the only seller in the market.

Natural Response

The natural response of a circuit refers to be behaviour of the circuit (in terms of voltage and current), in the absence of external excitation.

n-channel metal-oxide semiconductor (nMOS)

An n-channel metal-oxide semiconductor (nMOS) transistor is one in which n-type dopants are used in the gate region (the "channel"). A positive voltage on the gate turns the device on.

Negative

The opposite of positive. A potential less than that of another potential or of the earth. In electrical apparatus, the pole or direction toward which the current is suppose to flow.

Negative Feedback

Feeding a signal back to the input of an amplifier, or other circuit, that is proportional to the output signal, but having a phase that opposes the input signal.

Negative Sequence

A balanced set of three phase components which have the same magnitude, opposite sequence to the original unbalanced set, and phase angle diferring from each other by 120°. The frequency is of course the same as the original unbalanced three phase system.

Negative Terminal

The terminal of a battery from which electrons flow in the external circuit when the cell discharges.

Neighborhood EV (nEV)

NEVs operate below highway speeds—typically with a maximum speed of under 30mph. Not for highway use, but just fine for surface streets.

Net Capability

The maximum load-carrying ability of the equipment, exclusive of station use, under specified conditions for a given time interval, independent of the characteristics of the load. (Capability is determined by design characteristics, physical conditions, adequacy of prime mover, energy supply, and operating limitations such as cooling and circulating water supply and temperature, headwater and tailwater elevations, and electrical use.)

Net Generation

Gross generation minus plant use from all electric utility owned plants. The energy required for pumping at a pumped-storage plant is regarded as plant use and must be deducted from the gross generation.

Net Summer Capability

The steady hourly output, which generating equipment is expected to supply to system load exclusive of auxiliary power, as demonstrated by tests at the time of summer peak load.

Net Winter Capability

The steady hourly output which generating equipment is expected to supply to system load exclusive of auxiliary power, as demonstrated by tests at the time of winter peak demand.

Network Load

The designated load of a Transmission Customer, including the entire load of all Member Systems designated pursuant to Section 6.0. A Transmission Customer's Network Load shall not be reduced to reflect any portion of such load served by the output of any generating facilities owned, or generation purchased, by the Transmission Customer or its Member Systems.

Network Polymer

A polymer composed of trifunctional mer units that form three-dimensional molecules.

Network Synthesis

Finding a network that represents a given transfer function.

Neutral

Commonly, the return conductor in a circuit. It usually

has white insulation. More properly called the grounded conductor because it returns current to ground at the service panel. Note that this is different from the green-sheathed or bare copper grounding conductor that does not carry current except in case of equipment fault.

Neutral Conductor

A conductor connected to the neutral point of a system and contributing to the transmission of electrical energy. The term also means the equivalent conductor of an IT or d.c. system unless otherwise specified in the Regulations and also identifies either the mid wire of a three wire d.c. circuit or the earthed conductor of a two wire earthed d.c. circuit.

Neutral Ground Reactor

A reactor used to connect the neutral point of a three phase system to ground. Neutral Ground Reactors are used to limit ground fault current on Neutral Grounded (WYE) systems.

Neutral Grounding Resistor

A device that connects the neutral point of a three phase system to ground. Neutral Grounding Resistors are used to limit ground fault current on Neutral Grounded (WYE) systems.

Neutral Voltage Displacement (NVE)

A technique to measure the displacement of the neutral voltage with respect to earth.

Neutralizing Winding

An extra winding used to cancel harmonics developed in a saturated secondary winding, resulting in a sinusoidal output waveform from a ferroresonant transformer.

Neutron

A proton and an electron in very close union existing in the nucleus. A particle having the weight of a proton but carrying no electric charge. It is located in the nucleus of an atom.

New England Power Exchange (NEPEX)

This is the operating arm of the New England Power Pool.

New England Power Pool (NEPOOL)

A regional consortium of 98

utilities who coordinate, monitor and direct the operations of major generation and transmission facilities in New England.

Newton (N)

SI unit of force. One newton is equal to the force required to accelerate a body with the mass one kilogram by one metre per second per second.

Newton-Meter

Unit of torque, in the metric system, that is a force of one Newton, applied at a radius of one meter and in a direction perpendicular to the radius arm.

Nickel Cadmium Battery

The assembly of one or more cells with an alkaline electrolyte, a positive electrode of nickel oxide and negative electrodes of cadmium.

Nitrogen Oxides

Compounds of nitrogen and oxygen formed when fossil fuels burn.

Noct

Nominal Operating Cell Temperature. The solar cell temperature at a reference environment defined as 800 W/m2 irradiance, 20°C ambient air temperature, and 1 m/s wind speed with the cell or module in an electrically open circuit state.

Nodal Analysis

A method of analysis of circuits based on defining node voltages as the variables.

Node

point of connection between two or more elements or branches in a network.

Noise

An unwanted random signal (in the form of a voltage or current) in an electrical circuit making the information more difficult to identify.

Nominal

The normal operating value.

Nominal Capacity (Battery)

A designation by the battery manufacturer which helps identify a particular cell model and also provides an approximation of capacity. It is normally expressed in ampere-hours at a given discharge current.

Nominal Voltage

A nominal value assigned to a circuit or system for the

purpose of conveniently designating its voltage rating. The actual voltage at which a circuit operates can vary from the nominal within a range that permits satisfactory operation of equipment. A suitable approximate value of voltage used to designate or identify a System.

Non Return to Zero (NRZ)

A binary encoding scheme in which ones and zeroes are represented by opposite and alternating high and low voltages, and where there is no return to a zero (reference) voltage between encoded bits. That is, the stream has only two values: low and high.

Nonaqueous Batteries

Cells that do not contain water, such as those with molten salts or organic electrolytes.

Non-basic Service

Any category of service not related to basic services (generation, transmission, distribution and transition charges).

Non-bypassable Wires Charge

A charge generally placed on distribution services to recover utility costs incurred as a result of restructuring (stranded costs—usually associated with generation facilities and services) and not recoverable in other ways.

Noncoincidental Peak Load

The sum of two or more peakloads on individual systems that do not occur in the same time interval. Meaningful only when considering loads within a limited period of time, such as a day, week, month, a heating or cooling season, and usually for not more than 1 year.

Noncrystalline

The solid state wherein there is no long-range atomic order. Sometimes used synonymously with the terms amorphous, glassy and vitreous.

Non-cutoff (Lighting)

Luminaire light distribution is classified as non-cutoff when there is no candlepower limitation in the zone above max candlepower.

Non-Firm Power

Power or power producing capacity supplied or available under a commitment having limited or no assured availability.

Non-Firm Transmission Service

Point-to-point transmission service that is reserved and/or scheduled on an as-available basis and is subject to interruption. Non-firm Transmission Service is available on a stand-alone basis as either Hourly Non-firm Transmission Service or Short-Term Non-firm Transmission Service.

Non-Halogen Ethylene Copolymers

Non-Halogen Ethylene Copolymers combine attributes of polyethylene and polypropylene to produce cable insulating and jacketing compounds with superior fire protection. Unlike other ethylene compounds, these do not contain chemicals from the Halogen group of elements and do not give off toxic or acid byproducts when burned.

Non-inductive Circuit

A circuit in which the magnetic effect of the current flowing has been reduced by one several methods to a minimum or to zero.

Non-Inductive Load

A non-inductive load is a load in which the current is in phase with the voltage across the load.

Non-jurisdictional

Utilities, ratepayers and regulators (and impacts on those parties) other than state-regulated utilities, regulators and ratepayers in a jurisdiction considering restructuring. Examples include utilities in adjacent state and non-state regulated, publicly owned utilities within restructuring states.

Non-Linear

A characteristic which does not follow a straight line.

Non-Linear Load

A load where the wave shape of the steady state current does not follow the wave shape of the applied voltage.

Non-Loadbreak

Refers to a group of rubber insulating products that cannot be separated under load.

Nonrenewable Fuels

Fuels that cannot be easily made or "renewed." We can

use up nonrenewable fuels. Oil, natural gas, and coal are nonrenewable fuels.

Non-utility Generator

Independent power producers, exempt wholesale generators and other companies in the power generation business that have been exempted from traditional utility regulation.

Non-utility Power Producer

A corporation, person, agency, authority, or other legal entity or instrumentality that owns electric generating capacity and is not an electric utility. Nonutility power producers include qualifying cogenerators, qualifying small power producers, and other nonutility generators (including independent power producers) without a designated franchised service area, and which do not file forms listed in the Code of Federal Regulations, Title 18, Part 141.

Non-Volatile

Nonvolatile (NV) RAM is memory which retains its stored value when power is removed.

NOR Gate

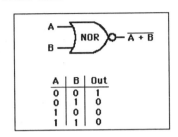

A	B	Out
0	0	1
0	1	0
1	0	0
1	1	0

A logic circuit that outputs a 1 only when each one of its inputs is a 0.

Normal Duty

A lightning impulse classifying current category for distribution class arresters defined by ANSI/IEEE C62.11. A normal duty rated arrester has 5000 amperage impulse value crest (refer to heavy duty).

Normal Operating Cell Temperature (NOCT)

The estimated temperature of a photovoltaic module when operating under 800 w/m2 irradiance, 20?C ambient temperature and wind speed of 1 meter per second. NOCT is used to estimate the nominal operating temperature of a module in its working environment.

Normal Temperature and Pressure (NTP)

Condition. Normal temperature may be taken as 0°C (physics) or 20°C (engineering) while normal pressure is taken as 760 torr.

Normally Open and Normally Closed

The terms "Normally Open" and "Normally Closed" are applied to a magnetically operated switching device (such as a contactor or relay, or to the contacts thereof) to signify the position taken when the operating magnet is de-energized.

Normal-Mode Noise

noise signal which appears between a set of phase conductors.

North/South

Technology factors are provided for North and South because some equipment and technologies are temperature sensitive. A North designation generally represents a utility that experiences cold winters and has average annual heating degree days of at least 5,000 (based on a 65 degree base). A South designation has relatively mild winters but a significant saturation of air conditioning. This geographical designation is very general, but it is intended to separate out areas that are warmer than others.

Norton's Theorem

States that a linear two-terminal circuit can be replaced by an equivalent circuit consisting of an equivalent current source and a shunt equivalent admittance.

NOT Circuit

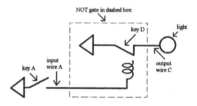

A logic circuit that inverts its only input.

Notching Relay

A relay which switches in response to a specific number of applied impulses.

Notice of Proposed Rulemaking

A designation used by the Federal Energy Regulatory Commission for some of its dockets.

Noxious Fumes

A combination of inert and corrosive gases usually associated with exhaust fumes or industrial by-products gases which can cause corrosive effects on temperature and pressure sensors when exposed.

N-Type

Negative semiconductor material in which there are more electrons than holes; current is carried through it by the flow of electrons.

N-Type Semiconductor

A semiconductor produced by doping an intrinsic semiconductor with an electron-donor impurity (e.g., phosphorous in silicon).

N-Type Silicon

Silicon material that has been doped with a material that has more electrons in its atomic structure than does silicon.

Nuclear Energy

Energy produced in the form of heat during the fission process in a nuclear reactor. When released in sufficient and controlled quantity, this heat energy may be used to produce steam to drive a turbine-generator and thus be converted to electrical energy.

Nuclear Fission

The splitting of an atomic nucleus, resulting in the release of large amounts of energy; the basic process a nuclear reactor uses to provide heat for the generation of electricity.

Nuclear Fuel

Fissionable materials that have been enriched to such a composition that, when placed in a nuclear reactor, will support a self-sustaining fission chain reaction, producing heat in a controlled manner for process use.

Nuclear Fusion

A nuclear reaction between light atomic nuclei as a result of which a heavier nucleus is formed and a large quantity of nuclear energy is released.

Nuclear Power

power released in exothermic (a reaction which gives off heat) nuclear reactions which can be converted to electric power by means of heat transformation equipment and a turbine-generator unit.

Nuclear Power Plant

A facility in which heat produced in a reactor by the fissioning of nuclear fuel is used to drive a steam turbine.

Nuclear Radiation

Invisible particles or waves given off by radioactive materials such as uranium.

Nuclear Regulatory Commission

This is the federal agency responsible for the licensing of nuclear facilities. They oversee these facilities and make sure regulations and standards are followed.

Nucleation

The initial stage in a phase transformation. It is evidenced by the formation of small particles (nuclei) of the new phase, which are capable of growing.

Nylon

For Wire and Cable applications, Nylon, a thermo-plastic compound, is used exclusively as a jacketing material. Nylon Jackets provide the insulation system a high degree of mechanical and chemical protection.

Nyquist

In A/D conversion, the Nyquist principle states that the sampling rate must be at least twice the maximum bandwidth of the analog signal in order to allow the signal to be completely represented. The maximum bandwidth of the signal (half the sampling rate) is called the Nyquist frequency (or Shannon sampling frequency).

In real life, sampling rate must be higher than that (because filters are not perfect). As an example, the bandwidth of a standard audio CD is a bit shy of the theoretical maximum of 22.05kHz (based on the sample rate of 44.1kHz).

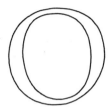

Obligation T Serve

The obligation of a utility to provide electric service to any consumer who seeks that service, and is willing to pay the rates set for that service. Traditionally, utilities have assumed the obligation to serve in return for an exclusive monopoly franchise.

Octahedral Position

The void space among closed-packed, hard sphere atoms or ions for which there are six nearest neighbors. An octahedron (double pyramid) is curcumscribed by lines constructed from centers of adjacent spheres.

Octal

A group of 8 symbols from 0 to 7.

Odd Symmetry Or Odd Function

A function has odd symmetry when its plot is anti-symmetrical about the vertical axis. $f(t) = -f(-t)$

Oerstead Oe

An old unit of magnetic field in the c.g.s. system. 1 Oe = 7.958×10^{-1} A/m

Off Grid

Anywhere not connected to the public utility power system.

Off Peak Power

Power supplied during designated periods of low power system demand.

Off-Line Uninterruptible Power Supply (UPS)

An uninterruptible power supply (UPS) which feeds power to the load directly from the utility and then transfers to battery power via an inverter after utility drops below a specified voltage. The delay between utility power loss and inverter startup can be long enough to disrupt the operation of some sensitive loads.

Off-Load Tap Changer

A tap changer that is not designed for operation while the transformer is supplying load.

Off-Peak Gas

Gas that is to be delivered and taken on demand when demand is not at its peak.

Off-Peak Power

Electricity supplied during periods of low system demand.

Off-Peak/On-Peak

Blocks of time when energy demand and price are matched as either high (on-peak) or low (off-peak).

Off-system sales

Sales by a utility to a customer (usually another utility) outside of its authorized market.

Ohm (Ω)

SI unit of electric resistance. One ohm is equal to the electric resistance between two points of a conductor when a constant potential difference of 1 V, applied to these points, produces in the conductor a current of 1 A, the conductor not being the seat of any electromotive force.

Ohm's Law

The formula that describes the amount of current flowing through a circuit. Ohm's Law—In a given electrical circuit, the amount of current in amperes (I) is equal to the pressure in volts (V) divided by the resistance, in ohms (R). Ohm's law can be shown by three different formulas: To find Current I = V/R To find Voltage V = I x R To find Resistance R = V/I

Ohmmeter

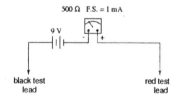

500 Ω F.S. = 1 mA

9 V

black test lead   red test lead

An instrument for measuring resistance in ohms. Take a look at this diagram to see how an ohmmeter is used to check a small control transformer. The ohmmeter's pointer deflection is controlled by the amount of battery current passing through the moving coil. Before measuring the resistance of an unknown resistor or electrical circuit, the ohmmeter must first be calibrated. If the value of resistance to be measured can be estimated within reasonable limits, a range selected that will give approximately half-scale deflection when the resistance is inserted between the probes. If the resistance is unknown, the selector switch is set on the highest scale. Whatever range is selected, the meter must be calibrated to read zero before the unknown resistance is measured.

Oil
A black liquid fossil fuel found deep in the Earth. Gasoline and most plastics are made from oil.

Oil Breakers
A type of high voltage circuit breaker using mineral oil as both an insulator and an interrupting medium. Typically, these units were produced for use at voltages from 35 kV to as much as 345 kV. Generally, these are older types and no longer produced for new installations.

Oil-Air (OA)
Oil-Air, a cooling classification for transformers now classified as ONAN. Oil type, Natural convection flow through cooling equipment and in windings, & Air external cooling medium.

Omnidirectional Antenna
This is like a dipole antenna because it radiates its signal 360 degrees horizontally; however, its signal is flatter than a dipole's allowing for higher gain.

Omni-polar
This type of device fully functions with either pole of a magnet.

On Load Tap Changer
A tap changer that can be operated while the transformer is supplying load.

One-Axis Tracking
A system capable of rotating about one axis.

On-line
A generating plant that is operating. When an operational plant is not on-line, it is "down."

Online (double conversion) UPS
A type of uninterruptible power supply (UPS) that is

constantly converting the incoming power from AC to DC and then inverting DC back to AC. This double conversion process prevents power disturbances from reaching protected loads. The UPS batteries will support the load if conditions exceed a set limit.

Online Uninterruptible Power Supply (UPS)

A UPS in which the inverter is on during normal operating conditions supplying conditioned power to the load through an inverter or converter that constantly controls the AC output of the UPS regardless of the utility line input. In the event of a utility power failure, there is no delay or transfer time to backup power.

On-Peak Energy

Energy supplied during periods of relatively high system demand as specified by the supplier.

Open Access

A regulatory mandate to allow others to use a utility's transmission and distribution facilities to move bulk power from one point to another on a nondiscriminatory basis for a cost-based fee.

Open Circuit

A circuit in which the flow of current is interrupted due to an open breaker or fuse. May be intentional or unintentional (as caused by a short).

Open Link

A fuse used on overhead electrical distribution systems that is held in place by two springs. This device and its holder have generally been replaced by Fused Cutouts where the fuse element in an arc tube.

Open-Circuit Voltage

The difference in potential between the terminals of a cell when the circuit is open (i.e., a no-load condition).

Open-Circuit Voltage (Battery)

The voltage of a cell or battery when it is not delivering or receiving power.

Open-Circuit Voltage

The maximum possible voltage across a photovoltaic cell or module; the voltage across the cell in sunlight when no current is flowing.

Opening Time

For a CB the time between energizing of the trip coil and the instant of contact parting. With a relay the operating time is defined as the time which elapses between the application of a characteristic quantity and the instant when the relay operates.

Operable Nuclear Unit

A nuclear unit is "operable" after it completes low power testing and is granted authorization to operate at full power. This occurs when it receives its full power amendment to its operating license from the Nuclear Regulatory Commission.

Operate AT (OAT)

The measured value, in AT, at which a reed contact closes. This is valid for the closing operation of form A, B, and E type reed contacts, and the change over operation from the normally closed contact to the normally open contact for form C and D type reed contacts.

Operate Time

The time interval from coil energization, to the closing of the reed contact. Where not otherwise stated, the functioning time of the reed contact in question is taken as its initial functioning time, not including contact bounce.

Operating Current

The current used by a lamp and ballast combination during normal operation.

Operating Point

The current and voltage that a photovoltaic module or array produces when connected to a load. The operating point is dependent on the load or the batteries connected to the output terminals of the array.

Operating Temperature

The range of temperature over which a device may be safely used. The temperature range which the device has been designed to operate.

Operating Time Characteristic

The curve depicting the relationship between different values of the characteristic quantity applied to a relay and the corresponding values of operating time.

Operating Value

The limiting value of the characteristic quantity at which the relay actually operates.

Operating Voltage

The value of the voltage under normal conditions at a given instant and at a given point in the system.

Operation and Maintenance Expenses

Costs that relate to the normal operating, maintenance and administrative activities of a business.

Operational Amplifier (Op amp)

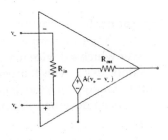

The ideal op amp is an amplifier with infinite input impedance, infinite open-loop gain, zero output impedance, infinite bandwidth, and zero noise. It has positive and negative inputs which allow circuits that use feedback to achieve a wide range of functions.

Using op amps, it's easy to make amplifiers, comparators, log amps, filters, oscillators, data converters, level translators, references, and more. Mathematical functions like addition, subtraction, multiplication, and integration can be easily accomplished.

Practical, real-world op amps have finite characteristics but in most applications, are close enough to the ideal to make a huge range of inexpensive, high-performance analog applications possible. They are the building block for analog design.

One key to op amp design is nodal analysis. Since the input impedance is infinite, the current in and out of the + and − input nodes defines the circuit's behavior.

Maxim has hundreds of op amps (and other amplifiers).

Operator

person handling equipment.

Optical Ground Wire (OPGW)

A ground wire that includes optical fibers to provide a communications link.

Optical Radiation

Electromagnetic radiation at wavelengths between the region of transition to Xrays (1 ? 1 nm) and the region of transition to radio waves (1 ??1 mm).

Options

Options are potential decisions over which a utility has a reasonable degree of control. One option might be to build a new coal-fired power plant; another option might be to refurbish an old power plant. Each option has one of more values to be specified. A specified option has a specified value such as year of implementation or size of plant. A plan is a set of specified options. A plan contains a set of decisions or commitments the utility can make, given the options available.

Optocoupler

A combination of an LED and a photodiode to give high isolation between the input and the output.

Optoelectronics

A technology that combines optics and electronics, including many devices based on the action of a pn junc-

tion. Examples are LEDs, photodiodes, and optocouplers.

OR Gate

A logic circuit that outputs a HIGH whenever one of its inputs is a HIGH.

Organic Light-Emitting Diode (OLED)

An LED made with organic materials. The diodes in displays made with OLEDs emit light when a voltage is applied to them. The pixel diodes are selectively turned on or off to form images on the screen. This kind of display can be brighter and more efficient than current LCD displays.

Orientation

Placement with respect to the cardinal directions, N, S, E, W; azimuth is the measure of orientation from north.

Origin of an Installation

The position at which electrical energy is delivered to an electrical installation.

Orthogonal Frequency Division Multiplexing (OFDM)

A method for multiplexing signals which divides the available bandwidth into a series of frequencies known as tones. Flarion uses the 5GHz channel and divides each channel into 400 discrete tones (each at slightly different frequency). Orthogonal tones do not interfere with each other when the peak of one tone corresponds with the null. All frequencies fade but the rapid switching, frequency-hopping technique is intended to allow more robust data service.

Oscillator

A circuit that produces an alternating waveform as output when the primarily powered by a direct input.

Oscillograph

An instrument for measuring alternating electric current or voltage by capturing the wave form. Electric Utilities use a variant called a Prefault Recorder, where the wave forms are stored for a short time on an ongoing basis and saved if the system sees an abnormal condition. These systems are also used to capture short duration transient conditions such as switching surges and lightning.

Oscilloscope

An instrument for making visible the instantaneous values of one or more rapidly varying electrical quantities as a function of time or of another electrical or mechanical quantity.

Other DSM Programs

A residual category to capture the effects of DSM programs that cannot be meaningfully included in any of the program categories listed and defined herein. The energy effects attributable to this category should be the net effects of all the residual programs. Programs that promote consumer's substitution of electricity by other energy types should be included in Other DSM Programs. Also, self-generation should be included in Other DSM Programs to the extent that it is not accounted for as backup generation in Other Load Management or Interruptible Load categories.

Outage

The state of a component or part of a power system that is not available for service because of some event associated with the component of power system. These are the longer term events (several seconds to hours) caused by external factors such as trees, car accidents, animals in contact with lines, operation of utility protective equipment, and weather conditions. Interruptions and long-term outages can cause costly lost production time.

Outage, Forced

1. An*outage that results from conditions directly associated with a power system component requiring that it be taken out of service either automatically or after switching operations can be performed. 2. An outage by improper operation of equipment or by human error.

Outage, Scheduled

An outage that results from intentionally taking a power system out of service, normally for maintenance or replacement.

Outlet

A point on the wiring system at which current is taken to supply utilization equipment.

Outlet Box

Medium-base incandescent lampholder designed for mounting in 31/4" or 4" electrical boxes. Available with or without pull-chain mechanism, and with or without built-in receptacle.

Output

The current, voltage, power, or driving force delivered by a circuit or device.

Output Capability

Expressed as capacity times voltage, or watt-hours.

Output common mode interface voltage

An unwanted alternating voltage which exists between each of the output terminals and a reference point.

Output Current of a transducer

The current produced by the transducer which is an analog function of the measurand.

Output Load

The total effective resistance of the circuits and apparatus connected externally across the output terminals.

Output Series Mode Interface Voltage

An unwanted alternating voltage appearing in series between the output terminals and the load.

Output Span

The algebraic difference between the lower and upper nominal values of the output signal.

Output to Input Ratio

The ratio between the sensed current and the output current of the amplifier.

Over current

Any current in excess of the rated current of equipment or the ampacity of a conductor. It may result from overload, short circuit or ground fault.

Over voltage

A voltage above the normal rated voltage or the maximum operating voltage of a device or circuit. A direct test over voltage is a voltage above the peak of the line alternating voltage.

Overcharge

Forcing current into a fully charged battery. The battery will be damaged if overcharged for a long period.

Overcurrent Detection

A method of establishing that the value of current in a circuit exceeds a predetermined value for a specified length of time.

Overcurrent Relay

A protection relay whose tripping decision is related to the degree by which the measured current exceeds a set value.

Over-drive (in AT)

The percentage of AT given above OAT, before measurement of CR.

Overload

Over-loading occurs when extra power is taken from the supply. The increased current due to over-loading will have an immediate effect on the cables; they will begin to heat up. If the over-loading is sustained the result could be an accelerated deterioration of the cable insulation and its eventual breakdown to cause an electrical fault.

Overload Current

An overcurrent occurring in a circuit which is electrically sound.

Overload Protection
Effect of a device operating on excessive current, but not necessarily on short circuit, to cause and maintain the interruption of current flow to the device being managed.

Overload Relay
A relay that responds to electric load and operates at a pre-set value of overload.

Overrange
The specified maximum operating point for which the stated accuracy condition applies.

Overshoot Time
The overshoot time is the difference between the operating time of the relay at a specified value of the input energizing quantity and the maximum duration of the value of input energizing quantity which, when suddenly reduced to a specific value below the operating level, is insufficient to cause operation.

Oversized
Wallplates that are approx. 3/4" higher and wider than the standard size and are used to conceal greater wall irregularities than those hidden by Midway wallplates. These wallplates are approx. 1/4" deep to ensure a proper fit when used with protruding devices.

Overvoltage
Similar to a surge but for a longer period of time, over 2.5 second.

Oxidation
A chemical reaction that results in the release of electrons by an electrode's active material.

Oxides of Nitrogen
Air emissions produced during the high temperature combustion of all fossil fuels. Oxides of nitrogen are chemically identified as NO, NO2 and NO3 or generically as NOx. In addition to contributing to ozone formation, some nitrogen oxides, such as NO2, are corrosive, acid rain precursors, and can cause respiratory problems. Act 141, Section 10r(3), requires all Michigan electric suppliers, beginning January 1, 2002, to disclose to customers the environmental characteristics of the average fuel mix used to produce the elec-

tricity products purchased by the customers. This disclosure will include the average air emissions, in pounds per megawatt-hour, of sulfur dioxide, carbon dioxide, and oxides of nitrogen.

Oxygen Recombination
The process by which oxygen generated at the positive plate during charge reacts with the pure lead material of the negative plate and in the presence of sulfuric acid and reforms water.

P/N

A semiconductor photovoltaic device structure in which the junction is formed between a p-type layer and an n-type layer.

Packet

When data is ready to be transmitted it is divided into pieces called packets. These packets contain information about which computer sent the data and where the data is going.

Packing Factor

The ratio of array area to actual land area or building envelope area, for a system; or, the ratio of total PV cell area to the total module area, for a module.

Pad Mounted Transformer

A transformer that is mounted on a pad (usually concrete or polycrete) that is used for underground service. Pad mounted transformers are available in single phase and three phase configurations.

Panel board

A single panel or group of panel units designed for assembly in the form of a single panel: includes buses and may come with or without switches and/or automatic over current protective devices for the control of light, heat, or power circuits of individual as well as aggregate capacity. It is designed to be placed in a cabinet or cutout box that is in or against a wall or partition and is accessible only from the front.

Paper

Normally consists of sheets of cellulose, mainly obtained from wood pulp from which lignin and other noncellulosic materials have been removed.

Parallel

Two or more elements are connected in parallel if they are connected to the same pair of nodes. parallel elements have the same voltage across them.

Parallel Connection

A way of joining two or more electricity-producing devices (i.e., PV cells or modules) by connecting positive leads together and negative leads together; such a configuration increases the current.

Parallel Interface

A parallel interface (as distinguished from a serial interface) is one in which data is sent on several wires (or several wireless channels) at once. Examples: GPIB, byte-wide parallel interfaces to data converters, memory and data buses on computer boards and backplanes.

In contrast, a serial interface uses one wire or wire-pair or wireless channel (or one in each direction).

Parallel Path Flow

This refers to the flow of electric power on an electric system's transmission facilities resulting from sched-uled electric power transfers between two other electric systems. (Electric power flows on all interconnected parallel paths in amounts inversely proportional to each path's resistance.)

Parallel Resonance

A resonance condition usually occurring in parallel RLC circuits, where the voltage becomes a maximum for a given current.

Paramagnetism

A relatively weak form of magnetism that results from the independent alignment of atomic dipoles (magnetic) with an applied magnetic field.

Parametric Conjunctive Test

A conjunctive test that ascertains the range of values of each parameter for which the test meets specific performance requirements.

Parasite Power

The device derives its supply power directly from the serial interface (1-Wire).

Parity Bit

An additional bit that is attached to each code group so that the total number of

1s being transmitted is either odd or even.

Parking Bushing

A bushing that is designed to accept a 200a elbow. Parking bushings are used to "Park" a hot cable that is terminated with a 200 Amp rated elbow.

Parking Stand

A metal bracket, usually made of steel, that is used to support a parking bushing that in turn is used to "Park" a medium voltage cable that is terminated with an 200 Amp rated elbow. Parking Stands are usually furnished mounted to the front panel of 200 Amp rated electrical apparatus such as pad mounted transformers and switchgear, on the covers of submersible transformers and on 200 Amp Junctions.

Parkway

Sometimes referred to as a rating for Grade Level Boxes or Covers rating.

Partial Discharge

Discharges which are partial and are not flash overs across electrodes.

Partial Load

An electrical demand that uses only part of the electrical power available.

Partial Zero Emissions Vehicle (PZEV)

PZEVs meet SULEV tailpipe emission standards, have zero evaporative emissions and a 15 year / 150,000 mile warranty. No evaporative emissions means that they have fewer emissions while being driven than a typical gasoline car has while just sitting.

Partition Locking

The ability to lockout writes and/or reads to certain sections of the memory.

Part-Winding—Starting

A part-winding start three-phase motor is one arranged for starting by first energizing part of its primary winding and, subsequently, energizing the remainder of the primary winding. The leads are normally numbered 1,2,3 (starting) and 7,8,9 (remaining).

Pascal (pa)

SI unit of pressure. One pascal is equal to the force of one newton exerted on one square meter.

Passivation

A chemical reaction that eliminates the detrimental effect of electrically reactive atoms on a photovoltaic cell's surface.

Passive Element

An element which does not generate electricity but either consumes it or stores it.

Passive Optical Network (PON)

A cost-effective way to provide high performance Fiber to the Home (FTTH) connectivity via shared optical fiber. PON connects up to 32 (or more) homes on the same network using passive optical components (splitters).

Passive Solar Home

A house that uses a room or another part of the building as a solar collector, as opposed to active solar, such as PV.

Pasted Plate (Battery)

Paste in which the active material is applied as a paste to a conductive grid.

Payback

The length of time it takes for the savings received to cover the cost of implementing the technology.

Payback Period

The length of time it takes for the savings received to cover the cost of implementing the technology.

p-Channel Metal-Oxide Semiconductor (pMOS)

A p-channel metal-oxide semiconductor (pMOS) transistor is one in which p-type dopants are used in the gate region (the "channel"). A negative voltage on the gate turns the device on.

Peak

The maximum instantaneous value of a varying voltage or current.

Peak Clipping

Peak clipping is used to reduce a utility's system peak, reducing the need to operate peaking units with relatively high fuel costs. Peak clipping is pursued only when the resources are not expected to be able to meet the impending load requirements.

Peak Current

The maximum value of an alternating current.

Peak Day

The day of highest customer demand for electricity during a year.

Peak Demand

The greatest demand placed on an electric system; measured in kilowatts or megawatts; also, the time of day or season of the year when that demand occurs.

Peak Factor

Ratio of the peak value to the r.m.s. value of an alternating waveform.

Peak Inverse Voltage

In an electron tube, the maximum negative voltage that can be applied to the plate without danger of arcover. In a semiconductor diode, the maximum reverse bias voltage that can be applied without reaching the zener (or breakdown) voltage.

Peak Load

The amount of electric power required by a consumer or a system during peak demand; measured in kilowatts or megawatts.

Peak Load Plant

A plant usually housing old, low-efficiency steam units, gas turbines, diesels, or pumped storage hydroelectric equipment normally used during the peak-load periods.

Peak Load Power plant

A power generating station that is normally used to produce extra electricity during peak load times.

Peak Load; Peak Demand

The maximum load, or usage, of electrical power occurring in a given period of time, typically a day.

Peak Power

Power generated by a utility unit that operates at a very low capacity factor; generally used to meet short-lived and variable high demand periods.

Peak Power Current

Amperes produced by a photovoltaic module or array operating at the voltage of the I-V curve that will produce maximum power from the module.

Peak Power Point

Operating point of the I-V (current-voltage) curve for a solar cell or photovoltaic module where the product of the current value times the voltage value is a maximum.

Peak Sun Hours

The equivalent number of hours per day when solar irradiance averages 1,000 w/m². For example, six peak sun hours means that the energy received during total daylight hours equals the energy that would have been received had the irradiance for six hours been 1,000 w/m².

Peak Surge Current

The maximum current allowed for a single impulse with continuous voltage applied.

Peak to Peak

The amplitude of the ac wave form from its positive peak to its negative peak.

Peak Value

The highest or maximum value of an alternation of alternating current or voltage. This peak value occurs twice during each cycle.

Peak Watt

A unit used to rate the performance of solar cells, modules, or arrays; the maximum

nominal output of a photovoltaic device, in watts (Wp) under standardized test conditions, usually 1,000 watts per square meter of sunlight with other conditions, such as temperature specified.

Peaking Capacity

Capacity of generating equipment normally reserved for operation during the hours of highest daily, weekly, or seasonal loads. Some generating equipment may be operated at certain times as peaking capacity and at other times to serve loads on an around-the-clock basis.

Peaking Plants

power plants that operate for a relatively small number of hours, usually during peak demand periods. Such plants usually have high operating costs and low capital costs.

Peaking Unit

A power generator used by a utility to produce extra electricity during peak load times.

Peak-to-Peak Value

The maximum voltage change occurring during

one cycle of alternating voltage or current. The total amount of voltage between the positive peak and the negative peak of one cycle or twice the peak value.

Pearlite

A two-phase microstructure found in some steels and cast irons. It results from the transformation of austenite of eutectoid compositions and consists of alternating layers of alpha-ferrite and cementite.

Pebi (pi)

Binary multiple prefix corresponding to petabinary or 2^{50} or $(2^{10})^5$ or 1024^5. [IEC 1998]

Pedestrian Loading

Refers to a grade level Reinforced Polymer Concrete or Fiberglass Reinforced Plastic Box or Cover loading applied by pedestrian traffic.

Pedestrian Walkway (Lighting)

A public walk for pedestrian traffic not necessarily within the right-of-way for vehicular traffic. Included are skywalks (pedestrian overpasses), subwalks (pedes-

trian tunnels), walkways giving access to parks or block interiors and mid-block street crossings.

Pelton Turbine

A type of impulse turbine with one or more jets of water hitting the buckets of a runner. The runner resembles a miniature water wheel. Pelton turbines are used in high head sites (20-6000 feet), and can be as large as 200 megawatts.

PEN Conductor

A conductor combining the functions of both protective conductor and neutral conductor.

Pendant

A type of switch designed for installation at the end of a length of portable cord or cable.

Penstock

A closed pipeline through which the water flows to a hydro turbine.

Percent Difference

The relative change in a quantity over a specified time period. It is calculated as follows: the current value has the previous value subtracted from it; this new number is divided by the absolute value of the previous value; then this new number is multiplied by 100.

Performance Attributes

Performance attributes measure the quality of service and operating efficiency. Loss of load probability, expected energy curtailment, and reserve margin are some of the performance attributes.

Periodic Function

A function which repeats itself after a definite period.

Permanent Magnet

An artificial magnet that retains its magnetism after the magnetizing force has been removed. Steel, when properly processed, can be made into a permanent magnet.

Permanent Magnet Alternator

An Alternator that uses moving permanent magnets instead of Electromagnets to induce current in coils of wire.

Permeability

A measure of how easily magnetic lines of force can pass through a material. The permeability of a material is defined as the constant of proportionality between the magnetic flux density and the magnetic field. It is a constant for a given magnetic material in the linear region.

Permeance

Inverse of reluctance. [Unit: H]

Permittivity

The permittivity of a material is defined as the constant of proportionality between the electric flux density and the electric field. It is a constant for a given dielectric.

Peta (P)

Decimal multiple prefix corresponding to 10^{15}.

Peukert's Law

Estimates the capacity of an electric battery over a range of discharge rates.

Phase

One of the characteristics of the electric service supplied or the equipment used. Practically all residential customers have single-phase service. Large commercial and industrial customers have either two-phase or three-phase service.

Phase Alternate Line (PAL)
A television standard used in most of Europe. Similar to NTSC, but uses subcarrier phase alternation to reduce the sensitivity to phase errors that would be displayed as color errors. Commonly used with 626-line, 50Hz scanning systems, with a subcarrier frequency of 4.43362MHz.

Phase Angle
The angular displacement between a current and voltage waveform, measured in degrees or radians.

Phase Angle Transducer
A transducer used for the measurement of the phase angle between two a.c. electrical quantities having the same frequency.

Phase Conductor
A conductor of an a.c. system for the transmission of electrical energy other than a neutral conductor, a protective conductor or a PEN conductor. The term also means the equivalent conductor of a d.c. system un-

less otherwise specified in the Regulations.

Phase Diagram
A graphical representation of the relationships between environmental constraints, composition, and regions of phase stability, ordinarily under conditions of equilibrium.

Phase Difference
Difference in phase angle between two sinusoids or phasors.

Phase Modulation
Modulation of the phase angle of a sinusoidal carrier by an amount proportional to the instantaneous value of the modulating wave.

Phase Relationship
The timing relationship between voltage and current. If voltage and current cross through zero in a cycle at the same time they are said to be in phase. Phase differences are expressed in degrees. A cycle is 360 degrees.

Phase Rotation
Phase rotation defines the rotation in a Poly-Phase System and is generally stated as "1-2-3", counterclockwise

rotation. Utilities in the United States use "A-B-C" to define there respective phase names in place "1-2-3". However some refer to there rotation as A-B-C, A-C-B, or C-B-A counterclockwise, were "A" can replace 1, 2, or 3. Europe adapted R-S-T to define the phase names.

Phase Sequence or Phase Rotation

The time order in which the voltages (or currents) pass through their respective maximum values (or any other definable position).

Phase Transformation

A change in the number and/ or character of the phases that constitute the microstructure of an alloy.

Phase Voltage

The voltage between the phase and neutral of a three phase system is defined as the phase-to-neutral voltage or more commonly as the phase voltage. For a balanced three phase system, the phase voltage is 1/S3 times the line-to-line voltage. The voltage across one arm of either a star-connected load or a delta-con-nected load is also sometimes referred to as the voltage across a phase or phase voltage. This latter quantity is the same as the phase-to-neutral voltage for a star-connected load, and the line-to-line voltage for a delta-connected load.

Phase-shift keying (PSK)

A modulation technique in which the phase of the carrier conveys the input signal's information.

Phasor

Representation of a sinusoid on the Argand diagram in the form of the magnitude (usually r.m.s.) and phase angle. It may be represented as a complex number in either cartesian co-ordinates or polar co-ordinates.

Phonon

A single quantum of vibrational or elastic energy.

Phosphor

A substance which is capable of luminescence. That is storing energy and later releasing it in the form of light.

Phosphorous (P)

A chemical element, atomic

number 15, used as a dopant in making n-semiconductor layers.

Photocell or Photo-Electric Cell
Device used for the detection and measurement of light.

Photocurrent
An electric current induced by radiant energy.

Photodiode
A reverse biased diode that is sensitive to incoming light.

Photoelectric
Descriptive of the effect which light has on electric circuits, through a device controlled by light.

Photoelectric Cell
A device for measuring light intensity that works by converting light falling on, or reach it, to electricity, and then measuring the current; used in photometers.

Photoelectrochemical Cell
A special kind of photovoltaic cell in which the electricity produced is used immediately within the cell to produce a useful chemical product, such as hydrogen. The product material is continuously withdrawn from the cell for direct use as a fuel or as an ingredient in making other chemicals, or it may be stored and used subsequently.

Photon
A quantum of electromagnetic radiation which has zero rest mass and energy equal to the product of the frequency of radiation and planck's constant.

Photovoltaic (PV)
Refers to any device which produces free electrons when exposed to light. When these electrons are properly gathered, a potential difference (voltage) may be produced by the device, for example, a solar cell produces approximately one half volt in full sun.

Photovoltaic (PV) Array
An interconnected system of PV modules that function as a single electricity-producing unit. The modules are assembled as a discrete structure, with common support or mounting. In smaller systems, an array can consist of a single module.

Photovoltaic (PV) Cell

The smallest semiconductor element within a PV module to perform the immediate conversion of light into electrical energy (direct current voltage and current). Also called a solar cell.

Photovoltaic (PV) Conversion Efficiency

The ratio of the electric power produced by a photovoltaic device to the power of the sunlight incident on the device.

Photovoltaic (PV) Device

A solid-state electrical device that converts light directly into direct current electricity of voltage-current characteristics that are a function of the characteristics of the light source and the materials in and design of the device. Solar photovoltaic devices are made of various semiconductor materials including silicon, cadmium sulfide, cadmium telluride, and gallium arsenide, and in single crystalline, multicrystalline, or amorphous forms.

Photovoltaic (PV) Effect

The phenomenon that occurs when photons, the "particles" in a beam of light, knock electrons loose from the atoms they strike. When this property of light is combined with the properties of semiconductors, electrons flow in one direction across a junction, setting up a voltage. With the addition of circuitry, current will flow and electric power will be available.

Photovoltaic (PV) Efficiency

The ratio of electric power produced by a cell at any instant to the power of the sunlight striking the cell. This is typically about 9% to 14% for commercially available cells.

Photovoltaic (PV) Generator

The total of all PV strings of a PV power supply system, which are electrically interconnected.

Photovoltaic (PV) Module

The smallest environmentally protected, essentially planar assembly of solar cells and ancillary parts, such as interconnections, terminals, [and protective devices such as diodes] intended to generate direct current power under unconcentrated sunlight. The structural (load carrying) member of a module can either be the top layer (superstrate) or the back layer (substrate).

Photovoltaic (PV) Panel

often used interchangeably with PV module (especially in one-module systems), but more accurately used to refer to a physically connected collection of modules (i.e., a laminate string of modules used to achieve a required voltage and current).

Photovoltaic (PV) Peak Watt

Maximum "rated" output of a cell, module, or system. Typical rating conditions are 0.645 watts per square inch (1000 watts per square meter) of sunlight, 68 degrees F (20 degrees C) ambient air temperature and 6.2 x 10-3 mi/s (1 m/s) wind speed.

Photovoltaics

Technology that produces electric power directly from the sunlight. A common application is in solar-powered pocket calculators, but various equipment remote from electric distribution lines also uses the technology.

Photovoltaic-Thermal (PV/T) system

A photovoltaic system that,

in addition to converting sunlight into electricity, collects the residual heat energy and delivers both heat and electricity in usable form. Also called a total energy system.

Physical Vapor Deposition

A method of depositing thin semiconductor films. With this method, physical processes, such as thermal evaporation or bombardment of ions, are used to deposit elemental semiconductor material on a substrate.

Pick up ratio

The ratio of the limiting values of the characteristic quantity at which the relay resets and operates. This value is sometimes called the differential relay.

Pick-up

A relay is said to 'pick-up' when it changes from the de-energized position to the energized position.

Pico (p)

Decimal sub-multiple prefix corresponding to one trillionth (US) or 10^{-12}.

Picocoulomb(s), (pC)

1. a unit of electrical charge.
2. PC: Printed circuit.
3. PC: Personal Computer.

Pid

A three mode control consisting of time Proportioning, Integral and Derivative rate action.

Piecewise Continuous Waveforms

Waveforms which are essentially continuous but in which multi-values occur over finite bounds. The waveform is single-valued and continuous in pieces.

Piezo Electric Effect

Vibration that occurs when a crystal is excited by an ac signal across its plates.

Piezoelectric

A dielectric material in which polarization is induced by the application of external forces.

Pigtail

A short length of wire attached to an existing wire or wires.

Pilot

A utility program offering a limited group of customers their choice of certified or licensed energy suppliers on a one year minimum trial basis.

Pilot Channel

A means of interconnecting between relaying points for the purpose of protection.

Pilot Light

A switch with an integral lamp in its actuator (toggle, rocker or pushbutton) that illuminates when the switch is connected to an energized circuit and the actuator is in the ON position.

Pilot Line

A cord or rope used to pull a heavier rope that will be used to pull a conductor into place.

P-I-N

A semiconductor device structure that layers an intrinsic semiconductor between a p-type semiconductor and an n-type semiconductor; this structure is most often used with amorphous silicon devices.

Pin and Sleeve

A connector with hollow, cylindrical sleeve-type contacts.

Planck's Constant

A universal constant that has a value of 6.63 ×10-34 J.

Planned Generator

A proposal by a company to install electric generating equipment at an existing or planned facility or site. The proposal is based on the owner having obtained (1) all environmental and regulatory approvals, (2) a signed contract for the electric energy, or (3) financial closure for the facility.

Plasma

An ionized gas containing about equal numbers of positive and negative charges, which is a good conductor of electricity, and is affected by a magnetic field.

Plastic

A solid material in the primary ingredient of which is an organic polymer of high molecular wight.

Plastic Deformation

Deformation that is permanent or nonrecoverable after release of the applied load.

Plasticizer

A low molecular weight polymer additive that enhances flexibility and workability and reduces stiffness and brittleness.

Plate

In an electron tube, the electrode that must be positive with respect to the cathode to allow the tube to conduct. Also called the Anode.

Plate (Battery)

The electrode of a cell consisting of a current collector and a positive or negative active material.

Plate Current

In an electron tube, the electron flow from cathode to plate. In the external plate circuit, the electron flow from the plate through the circuit, to the cathode.

Plate Voltage

In an electron tube, the difference of potential between plate and cathode.

Plates

In a capacitor, the plates are the conducting surfaces.

Platinum Resistance Thermometer (PRT)

A resistance temperature device (RTD).

Plenum

Chamber or space forming a part of an air conditioning system

Plug

A device, provided with contact pins, which is intended to be attached to a flexible cable, and which can be engaged with a socket outlet or with a connector.

Plug Setting Multiple

A term used in conjunction with electromechanical relays, denoting the ratio of the fault current setting of the relay.

PN Junction

A junction between an N-type semiconductor and a P-type semiconductor made by some method of diffusing, fusing or melting.

Pocket Current Transformer

A round or toroidal core transformer mounted on bushings of power transformers, bulk oil circuit breaker, and other dead tank circuit breakers. These transformers are placed in pockets of these elements they are mounted on where the pocket length is measured in inches.

Pocket Plate

A plate for a battery in which active materials are held in a perforated metal pocket.

Point (In Wiring)

A termination of the fixed wiring intended for the connection of current using equipment.

Point Defect

A crystalline defect associated with one or, at most, several atomic sites.

Point of Common Coupling (PCC)

The location of the connection between the CEB network and the Embedded Generator, beyond which other customer loads may be connected on the CEB side. The PCC may be separate from the Point of Supply where a line is dedicated to the connection of an Embedded Generator.

Point of Supply (POS)

The location of the connection between the CEB network and the Embedded Generator.

Point(s) of Delivery

Point(s) of interconnection on the Transmission

Provider's Transmission System where capacity and/or energy transmitted by the Transmission Provider will be made available to the Receiving Party. The Point(s) of Delivery shall be specified in the Service Agreement.

Point(s) of Receipt

Point(s) of interconnection on the Transmission Provider's Transmission System where capacity and/or energy will be made available to the Transmission Provider by the Delivering Party. The Point(s) of Delivery shall be specified in the Service Agreement.

Point-Contact Cell

A high efficiency silicon concentrator cell that employs light trapping techniques and point-diffused contacts on the rear surface for current collection.

Point-on-Wave (POW)

Point-on-wave switching is the process to control moment of switching to minimize the effects (inrush currents, over-voltages).

Point-to-Point Transmission Service

The reservation and/or transmission of energy on either a firm basis and/or a non-firm basis from Point(s) of Receipt to Point(s) of Delivery, including any Ancillary Services that are provided by the Transmission Provider in conjunction with such service.

Point-to-Point Transmission Service Tariff

The Transmission Provider's Point-to-Point Transmission Service Tariff as such tariff may be amended and/or superseded from time to time.

Poisson's Ratio

For elastic deformation, the negative ratio of lateral and axial strains that result from an applied axial stress.

Polar Molecule

A molecule in which there exists a permanent electric dipole moment by virtue of the asymmetrical distribution of positively and negatively charged regions.

Polarisation

Change in the potential of an electrode as the result flow.

Polarity

1) The electrical Term used to denote the voltage rela-

tionship to a reference potential (+). 2) With regard to Transformers, Polarity is the indication of the direction of the current flow through the high voltage terminals with respect to the direction through the low voltage terminals.

Polarization (electronic)
For an atom, the displacement of the center of the negatively charged electron cloud relative to the positive nucleus, which is induced by an electric field.

Polarization (ionic)
Polarization as a result of the displacement of anions and cations in opposite directions.

Polarization (orientation)
Polarization resulting from the alignment (by rotation) of permanent electric dipole moments with an applied electric field.

Polarized
A system in which the slots/ blades for the hot leads are narrower than those for the neutral leads.

Poles
The magnetic poles set up

inside an electric machine by the placement and connection of the windings.

Polycrystalline
Referring to crystalline materials that are composed of more than one crystal or grain.

Polycrystalline Silicon
A material used to make photovoltaic cells, which consist of many crystals unlike single-crystal silicon.

Polyethylene
A thermoplastic material composed of ethylene polymers. Polyethylene has excellent electrical and mechanical properties and is used an insulating material in cable.

Polygon of Forces
If the forces acting on an object can be represented by the sides of a polygon taken in order, the forces will be in equilibrium.

Polymer
A solid, nonmetallic (normally organic) compound of high molecular weight the structure of which is composed of small repeat (or mer) units.

Polymer Concrete

Also referred to as Reinforced Polymer Mortar (RPM). Polymer Concrete material consists of calcareous and siliceous stone, glass fibers and thermoses polyester resin. Polymer concrete can be used in the manufacture of equipment pads, grade level boxes and box covers.

Polymorphism

The ability of a solid material to exist in more than one form or crystal structure.

Polyphase

A polyphase system is a means of distributing alternating current electrical power. Polyphase systems have two or more energized electrical conductors carrying alternating currents with a definite time offset between the peak amplitudes of the wave in each conductor. In modern utility power generation and distribution three phases are used, with the phases separated in time by one third of an AC cycle. A polyphase system must provide a defined direction of phase rotation, so mirror image voltages do not count towards the phase order. A 3-wire system with two phase conductors 180 degrees apart is still only single phase. Such systems are sometimes described as split phase.

Porcelain

Hard, white material made by the firing of a mixture of pure kaolin (china clay) with felspar and quartz, or with other materials containing silica.

Positive

The term used to describe a terminal with fewer electrons than normal so that it attracts electrons. Electrons flow into the positive terminals of a voltage source.

Positive Feedback

Feedback where the returning signal aids the effect of the input signal.

Positive Sequence

A balanced set of three phase components which have the same magnitude, same sequence as the original unbalanced set, and phase angle deferring from each other by 120°. The frequency is of course the same as the original unbalanced three phase system.

Positive Temperature Coefficient (PTC)

When the resistance of a component rises with temperature, it is said to have a positive temperature coefficient.

Example: Hewlett-Packard's first commercial product, an audio oscillator, used a common light bulb as a PTC element in the feedback circuit to maintain constant output amplitude regardless of frequency.

Positive Terminal

The terminal of a battery toward which electrons flow through the external circuit when the cell discharges.

Pot

1) Slang for an overhead transformer. 2) Short for "Potential".

Potential

The voltage in a circuit. Reference is usually to the AC Voltage.

Potential Difference

potential difference is the work done in moving a unit positive electric charge from one point to another. [Unit: volt or V]

Potential Divider

A combination of impedances which allows a fraction of the input voltage to be taken as output.

Potential Energy

Energy in a stored form. Batteries store potential electrical energy. Water behind a dam also had potential energy, because the stored water can be released for future power production.

Potential Equalization Conductor

Conductor providing a connection between equipment and the potential equalization busbar of the electrical installation.

Potential Peak Load Reduction

The amount of annual peak load reduction capability (measured in kilowatts) that can be deployed from Direct Load Control, Interruptible Load, Other Load Management, and Other DSM Program activities. It represents the load that can be reduced either by the direct control of the utility system operator or by the consumer in response to a utility request to curtail load. It reflects the installed load reduction capability, as opposed to the Actual Peak Reduction achieved by participants, during the time of annual system peak load.

Potential Transformer

A transformer used measure the amount of Voltage in a circuit. Its primary is rated in excess of the expected voltage of the circuit and the secondary will normally be rated at 120 volts being equal to the nominal full primary voltage.

Potentiometer

A three terminal device with a wiper that is positioned along a resistive element; effectively making a voltage divider.

Pothead

Slang for a device used to transition an overhead conductor to underground. Potheads are normally porcelain and have been largely replaced with non-ceramic, synthetic rubber, terminators of the type manufacturered by Thomas & Betts Elastimold.

Pound (lb)

The imperial unit of weight. 1 lb = 0.453,592 kg

Poundal

The imperial unit of force. It is that force which acting on a mass of 1 lb will impart to it an acceleration of 1 ft/s^2.

Pound-Foot

Unit of torque, in the English system, that is a force of one pound, applied at a radius of one foot, and in a direction perpendicular to the radius arm.

Power

The rate at which work is done; it is usually expressed as the number of foot pounds in one minute, that is, if you lift 33,000 foot pounds in one minute, you have done 1 horsepower of work.

Power Added Efficiency

In an RF power amplifier, power added efficiency (PAE) is defined as the ratio of the difference of the output and input signal power to the DC power consumed.

Power amplifier (PA)

An amplifier used to drive significant power levels. An audio amplifier that drives a loudspeaker and the final stage of a transmitter are common examples.

Power Conditioning

The process of modifying the characteristics of electrical power (for e.g., inverting direct current to alternating current).

Power Conditioning Equipment

Electrical equipment, or power electronics, used to convert power from a photovoltaic array into a form suitable for subsequent use. A collective term for inverter, converter, battery charge regulator, and blocking diode.

Power Conditioning Systems

A broad class of equipment that includes filters, isolation transformers, and voltage regulators. Generally, these types of equipment offer no protection against power outages.

Power Conversion Efficiency

The ratio of output power to input power of the inverter.

Power Density

The ratio of the power available from a battery to its mass (W/kg) or volume (W/l).

Power Dissipation

The amount of power that is consumed and converted to heat.

Power Distribution Unit

A portable electrical distribution unit that provides an easily expandable and flexible electrical environment for an equipment and its associated peripherals.

Power Electronics Device

An electronic device (e.g. thyristor or IGBT) or assembly of such devices (e.g. inverter). Typically used in a power transmission system to provide smooth control of output of an item of plant.

Power Exchange

The entity that will establish

a competitive spot market for electric power through day- and/or hour-ahead auction of generation and demand bids.

Power Exchange Generation
Generation being scheduled by the power exchange.

Power Exchange Load
Load that has been scheduled by the power exchange and which is received through the use of transmission or distribution facilities owned by participating transmission owners.

Power Factor
The ratio of energy consumed (watts) versus the product of input voltage (volts) times input current (amps). In other words, power factor is the percentage of energy used compared to the energy flowing through the wires. Adding capacitors to the system changes the inductive effect of the ballast coils, converting a Normal Power Factor (NPF) to a High Power Factor (HPF) system.

Power Factor Correction
Process of increasing the power factor to near unity without altering the original load.

Power Factor Correction Capacitor
This is a device that helps improve the efficiency of the flow of electricity through distribution lines by reducing energy losses. It is installed in substations and on poles. Usually it is installed to correct an unwanted condition in an electrical system.

Power Fail
A feature in a microprocessor supervisory circuit that provides early warning to the microprocessor of imminent power failure.

Power Grid
A network of power lines and associated equipment used to transmit and distribute electricity over a geographic area.

Power Line Carrier Communication
A mean of transmitting information over a power transmission line by using a carrier frequency superimposed on the normal power frequency.

Power Management Integrated Circuit (PMIC)

Circuits used to regulate and control power.

Power Marketer

An entity that takes title to electric power and then re-sells the power to end-use customers. This "middleman," which acts for itself in negotiating contracts, purchases, or sales of electrical energy, is required to meet two FERC tests to be certified as a power marketer: 1) show lack or adequate mitigation of transmission power; and 2) prove non-dominance of market power.

Power Marketers

Business entities engaged in buying, selling, and marketing electricity. Power marketers do not usually own generating or transmission facilities. Power marketers, as opposed to brokers, take ownership of the electricity and are involved in interstate trade. These entities file with the Federal Energy Regulatory Commission for status as a power marketer.

Power Outage

An interruption of power.

Power Outlet

An enclosed assembly that may include receptacles, circuit breakers, fuseholders, fused switches, buses, and watt-hour meter mounting means; intended to supply and control power to mobile homes, recreational vehicles, park trailers, or boats; or to serve as a means for distributing power required to operate mobile or temporarily installed equipment.

Power Plant

A generating station where electricity is produced.

Power Pool

An association of two or more interconnected electric systems having an agreement to coordinate operations and planning for improved reliability and efficiencies.

Power Purchase Agreement

A contract entered into by an independent power producer and an electric utility. The power purchase agreement specifies the terms and conditions under which electric power will be generated and purchased. Power purchase agreements generally include: specification of the size and operating parameters of the generation facility; contract terms; price mechanisms; service and performance obligations; dispatchability options; and conditions of termination or default.

Power Supply

The part of a circuit that supplies power to the entire circuit or part of the circuit. Usually a separate unit that supplies power to a specific part of the circuit in a system.

Power Supply Rejection Ratio (PSRR)

The ability of an amplifier to maintain its output voltage as its power-supply voltage is varied.

Power Surge

A sudden, sharp increase in the voltage or current lasting less than one cycle.

Power Transformer

A large transformer, generally larger than 1,000 kVA in capacity.

PowerBack Contractors

Electrical contractors certified by PGE that are available within 24 hours of receiving a call to restore a customer's electrical service in the event of damage. A Power Back Contractor can be reached 24 hours a day, seven days a week to schedule home electrical repairs.

PowerCap

A special surface-mount package with access to the internal cavity via an openable top. This packaging scheme allows easy upgrade of NV RAMs without having to change the PCB hardware layout. The user can simply open the lid and swap out the IC.

Power-over-Ethernet (PoE)

A means for delivering power to a remote device using the same cable lines used to deliver Ethernet data.

Pre-Bias Soft Start

A power-supply feature that prevents discharging of the output capacitor when the power supply starts up. Discharging the output capacitor could create either start up oscillation problems at cold start or large voltage disturbances on the output voltage bus at hot plug-in. Pre-bias soft start is an important feature in redundant power-supply systems, parallel power supply modules, battery back-up voltage buses, and other applications where multiple power sources supply one node.

Precipitation Hardening

Hardening and strengthening of a metal alloy by extremely small and uniformly dispersed particles that precipitate from a supersaturated solid solution.

Precision

Degree of agreement within a group of measurements or instruments.

Preemphasis

In some transmission and recording systems (e.g. vinyl records, FM radio, analog magnetic tape), there is more noise at higher frequencies. To offset this, the audio signal is "preemphasized" at the transmitter — filtered with a high-pass filter to boost the higher audio frequencies. A matching low-pass filter is used at the receiver to return to an overall flat audio-frequency response. The filter at the receiver reduces the high-frequency noise intro-

duced by the transmission process.

Prepreg

Continuous fiber reinforcement pre-impregnated with a polymer resin which is then partially cured.

Present Value

The amount of money required to secure a specified cash flow on a future date at a given rate of return.

Present Worth Factor

The adjustment factor that discounts a sum of future dollars back to the current year.

Pressure

The force per unit area acting on a surface.

Pressure Cooker Test

A Pressure Cooker Test (PCT) tests a part under high temperature, humidity, and pressure conditions. Also called an Autoclave Test or Pressure Pot Test (PPOT).

Primary Bond

Interatomic bonds that are relatively strong and for which bonding energies are relatively large.

Primary Cell

A cell designed to produce electric current through an electrochemical reaction that is not efficiently reversible. The cell, when discharged, cannot be efficiently recharged by an electric current. Alakline, lithium, and zinc air are common types of primary cells.

Primary Circuit

Distribution circuit (less than 69,000 volts) on the high voltage side of the transformer.

Primary Voltage Rating

Designates the input circuit voltage for which the primary winding is designed.

Prime Mover

The engine, turbine, water wheel, or similar machine that drives an electric generator; or, for reporting purposes, a device that converts energy to electricity directly (e.g., photovoltaic solar and fuel cells).

Printed Circuit

A method of manufacturing parts of electronic equipment in which the wiring between components, and certain fixed components themselves, are printed on to an insulating board.

Printed Circuit Board

A printed circuit board, or PC board, or PDB, is a nonconductive material with conductive lines printed or etched. Electronic components are mounted on the board and the traces connect the components together to form a working circuit or assembly.

A PC board can have conductors on one side or two sides and can be multilayer — a sandwich with many layers of conductors, each separated by insulating layers.

The most common circuit boards are made of plastic or glass-fiber and resin composites and use copper traces, but a wide variety of other materials may be used. Most PCBs are flat and rigid but flexible substrates can allow boards to fit in convoluted spaces.

Components are mounted via SMD (surface-mount) or through-hole methods.

Private Mobile Radio (PMR)

Radio bands generally for use within a defined user

group, such as the emergency services or by the employees of a mining project.

Probe

An oscilloscope input device, usually having a pointed metal tip for making electrical contact with a circuit element and a flexible cable for transmitting the signal to the oscilloscope.

Process Failure Mode and Effects Analysis (PFMEA)

A methodology for assessing the weaknesses of production processes and the potential effects of process failures on the product being produced.

Process Heating

Energy Efficiency program promotion of increased electric energy efficiency applications in industrial process heating.

Production Part Approval Process (PPAP)

Used by automotive industry for acceptance of new products for release and use on automobiles.

Propeller

The spinning thing that makes an airplane move forward. Often incorrectly used to describe a wind turbine Rotor.

Propeller Turbine

A type of reaction turbine with a propeller-style runner. Water passes through the runner and drives the propeller blades. Propeller turbines can be used from 2 to 300 feet of head, and can be as large as 100 megawatts.

Proportional Limit

The point on a stress-strain curve at which the straight line proportionality between stress and strain ceases.

p-type Semiconductor

A semiconductor for which the predominant charge carriers responsible for electrical conduction are holes. Normally, acceptor impurity atoms give rise to the excess holes.

Prospective Fault Current

The value of overcurrent at a given point in a circuit resulting from a fault of negligible impedance between live conductors having a difference of potential under normal operating conditions, or between a live conductor and an exposed conductive part.

Protected Zone

The portion of a power system protected by a given protection system or a part of that protection system.

Protection Equipment

The apparatus, including protection relay, transformers and ancillary equipment, for use in a protection system.

Protection Relay

A relay designed to initiate disconnection of a part of an electrical installation or to a warning signal, in the case of a fault or other abnormal condition in the installation. A protection relay may include more than one electrical element and accessories.

Protection System

A combination of protection equipment designed to secure, under pre-determined conditions, usually abnormal, the disconnection of an element of a power system, or to give an alarm signal, or both.

Protective Conductor Current

Electric current which flows in a prospective conductor under normal operating conditions:

Protective Earth Conductor

Conductor to be connected between the protective earth terminal and an external protective earthing system.

Protective Earth Terminal

Terminal connected to conductive parts of Class I equipment for safety purposes. This terminal is intended to be connected to an external earthing system by a protective earth conductor.

Protective Extra Low Voltage (PELV)

An extra low voltage system which is not electrically separated from earth, but which otherwise satisfies all the requirements for SELV.

Protective Multiple Earthing (PME)

An earthing arrangement, found in TN C S systems, in which the supply neutral conductor is used to connect the earthing conductor of an installation with Earth, in accordance with the Electricity Safety, Quality and Continuity Regulations 2002.

Protector

A protector is another name for an arrester or diverter.

Proton

The smallest quantity of electricity which can exist in the free state. A positive charged particle in the nucleus of an atom.

Proximity Sensor or Proximity Switch

A sensor or switch with the ability to detect it's relationship to a metal target without making physical contact.

Proxy Server

A system that caches items from other servers to speed up access. On the web, a proxy first attempts to find data locally, and if it is not available, obtains it from the remote server where the data resides permanently.

psi

Pounds per square inch

psi lb/in²

Unit of pressure in the imperial system. 1 psi = 6895 pa

P-Type Semiconductor

A semiconductor in which holes carry the current; produced by doping an intrinsic semiconductor with an electron acceptor impurity (e.g., boron in silicon).

Public Street and Highway Lighting

Public street and highway lighting includes electricity supplied and services rendered for the purposes of lighting streets, highways, parks, and other public places; or for traffic or other signal system service, for municipalities, or other divisions or agencies of State or Federal governments.

Pull

A noun referring to the installation of one or more cables.

Pull Tension

The tension in pounds or kilograms required to pull a cable or wire into a duct or conduit or into an overhead location.

Pull-Chain

An incandescent lampholder with an internal

switching mechanism that is activated by pulling down on a beaded chain or cord.

Pulling

The act of installing one or more cables.

Pulse-Code Modulation (PCM)

Signal encoding through modulation of a carrier pulse width; a signal encoding technique used in digital audio and some switching power supplies.

Pulse-Width Modulation (PWM) Temperature Sensor

Temperature sensor with digital, logic-level output. The output has a fixed frequency and the duty cycle varies with the measured temperature.

Pulse-Width-Modulated (PWM) Wave Inverter

A type of power inverter that produce a high quality (nearly sinusoidal) voltage, at minimum current harmonics.

Pumped Storage

A facility designed to generate electric power during peak load periods with a hydroelectric plant using water pumped into a storage reservoir during off-peak periods.

Pumped Storage Hydroelectric Plant

A plant that usually generates electric energy during peak-load periods by using water previously pumped into an elevated storage reservoir during off-peak periods when excess generating capacity is available to do so. When additional generating capacity is needed, the water can be released from the reservoir through a conduit to turbine generators located in a power plant at a lower level.

Puncture

Term used when a disruptive discharge occurs through a solid dielectric. A disruptive discharge in a solid dielectric produces a permanent loss of dielectric strength; in a liquid of gaseous dielectric, the loss may be only temporary.

Purchased Power Adjustment

A clause in a rate schedule that provides for adjustments to the bill when energy from another electric system is acquired and it varies from a specified unit base amount.

Pure Pumped Storage Hydro-electric Plant

A plant that produces power only from water that has previously been pumped to an upper reservoir.

Pure Pumped-Storage Hydro-electric Plant

A plant that produces power only from water that has previously been pumped to an upper reservoir.

Pushbutton

Part of an electrical device, consisting of a button that must be pressed to effect an operation.

Pyrheliometer

An instrument used for measuring direct beam solar irradiance. Uses an aperture of 5.7° to transcribe the solar disc.

Pyronometer

An instrument for measuring total hemispherical solar irradiance on a flat surface, or "global" irradiance; thermopile sensors have been generally identified as pyranometers, however, silicon sensors are also referred to as pyranometers.

Quad

A measure of energy equal to one trillion Btus; an energy equivalent to approximately 172 million barrels of oil.

Quadrant

Quarter circle. Sector of a circle bounded by an arc and two radii at right angles.

Quadratic Equation

An equation involving terms up to the second power of the unknown quantity. It has two roots which satisfy the equation.

Quadrature

The relation between two waves of the same frequency, but one-quarter of a cycle (90°) out of phase.

Quadrature Amplitude Modulation (QAM)

A modulation method in which two signals are used to amplitude-modulate two carriers that are in quadra-ture (90 degrees out of phase with each other). The two modulated signals are combined.

A common application is in PAL and NTSC color television transmission. Color is encoded into two analog signals (called I and Q), which modulate quadrature color carriers.

Modems also use this approach, to increase the data bandwidth they can carry (or, more accurately, to trade bandwidth for error rate or noise immunity).

Quadrature Phase Shift Keying (QPSK)

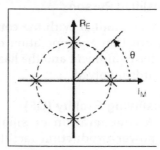

Quadrature Phase Shift Keying (QPSK) is a form of Phase Shift Keying in which two bits are modulated at once, selecting one of four possible carrier phase shifts (0, 90, 180, or 270 degrees). QPSK allows the signal to carry twice as much information as ordinary PSK using the same bandwidth. QPSK is used for satellite transmission of MPEG2 video, cable modems, videoconferencing, cellular phone systems, and other forms of digital communication over an RF carrier.

Qualification Test

A procedure applied to a selected set of photovoltaic modules involving the application of defined electrical, mechanical, or thermal stress in a prescribed manner and amount. Test results are subject to a list of defined requirements.

Qualified Person

One familiar with the construction and operation of the equipment and the hazards involved.

Qualifying Facility (QF)

A cogeneration or small power production facility that meets certain ownership, operating, and efficiency criteria established by the Federal Energy Regulatory Commission (FERC) pursuant to the Public Utility Regulatory Policies Act (PURPA).

Quality Control

Inspection, analysis and action required to ensure quality of output.

Quality Factor (Q)

The quality factor of a resonant circuit is the ratio of its resonant frequency to its bandwidth. The higher the quality factor, the lower the losses in the circuit and higher the resonant peak.

Quantisation Error

The error caused by nonzero resolution of an analog to digital converter. It is an inherent error of the device.

Quantization

A process whereby the continuous range of input-signal values is divided into nonoverlapping subranges. Each of these subranges has a discrete value of the output uniquely assigned. Once a signal value falls within a given subrange, the output provides the corresponding discrete value.

Quantum

One of the very small discrete packets into which many forms of energy are subdivided.

Quantum Electronics

Applying molecular physics to electronics.

Quark

A hypothetical basic subatomic nuclear particle believed to be the basic component of protons, neutrons, etc.

Quartz

A form of silicone dioxide. Commonly used in the making of radio transmitters and heat resistant products.

Quick Disconnect

A type of connector shell that permits rapid locking and unlocking of two connector halves.

Quick Disconnect Coupling

A design feature, apparent in the quick disconnect connector; it permits relatively rapid joining and separation.

Quick-Break

A switch or circuit breaker that has a high contact opening speed.

Quiescent

For an electronic circuit, a quiet state in which the circuit is driving no load and its inputs are not cycling. Most commonly used for the specification "quiescent current," the current consumed by a circuit when it in a quiescent state.

Rad

The unit of absorbed dose of ionising radiation. One rad is equal to the energy absorption of 100 ergs per gram of irradiated material.

Radar

An acronym for Radio Detection And Ranging. A general term now used to include any system employing microwaves for the purpose of locating, identifying, navigating, or guiding such moving objects as ships, aircraft, etc.

Radial Circuits

Each circuit commences from the consumer unit/distribution board through an MCB/fuse of specific rating (e.g. 20A), loop into each socket outlet but ends at a socket outlet (does not return to the original fuse/mcb). No loops are formed.

Radian (Rad)

The radian is SI unit of the plane angle. It is a supplementary unit. It is defined as the plane angle between two radii of a circle that cut off on the circumference an arc equal in length to the radius.

Radiant Efficiency

Ratio of the radiant flux (power) emitted from a source to the power consumed.

Radiation

RF energy which is emitted or leaks from a distribution system and travels through space. These signals often cause interference with other communication services.

Radio

The use of electromagnetic radiation to communicate electrical signals without wires. In the widest sense the term incorporates sound broadcasting, television and radar.

Radio Frequency (RF)
An AC signal of high enough frequency to be used for wireless communications.

Rain tight
So constructed or protected that exposure to a beating rain will not result in the entrance of water.

Rainproof
So constructed, projected, or treated as to prevent rain from interfering with the successful operation of the apparatus under specified test conditions.

Ramp Rate
The rate at which you can increase load on a power plant. The ramp rate for a hydroelectric facility may be dependent on how rapidly water surface elevation on the river changes.

Ramp Up (Supply Side)
Increasing load on a generating unit at a rate called the ramp rate.

Ramp-Up (Demand-Side)
Implementing a demand-side management program over time until the program is considered fully installed.

Random Jitter
Random jitter (RJ) includes all jitter components not defined as deterministic jitter (i.e., the jitter that is not related to the signal and known noise sources).

Range
Nominal operating limits, specified by the lowest calibration point to the highest calibration point.

Rankine ^{o}r Scale of Temperature
It is the absolute Fahrenheit scale, starting from absolute zero at −459.69 °F. Can be converted to degree Fahrenheit by the addition of 459.69 °F.

Rate Base
The value of property upon which a utility is permitted to earn a specified rate of return as established by a regulatory authority. The rate base generally represents the value of property used by the utility in providing service and may be calculated by any one or a combination of the follow-

ing accounting methods fair value, prudent investment, reproduction cost, or original cost. Depending on which method is used, the rate base includes cash, working capital, materials and supplies, and deductions for accumulated provisions for depreciation, contributions in aid of construction, customer advances for construction, accumulated deferred income taxes, and accumulated deferred investment tax credits.

Rate Class
A group of customers identified as a class and subject to a rate different from the rates of other groups.

Rate Structure
The design and organization of billing charges by customer class to distribute the revenue requirement among customer classes and rating period.

Rate-Basing
The practice by utilities of allotting funds invested in utility Research Development Demonstration and Commercialization and other programs from ratepayers, as opposed to allocating these costs to shareholders.

Rated Battery Capacity
The term used by battery manufacturers to indicate the maximum amount of energy that can be withdrawn from a battery under specified discharge rate and temperature.

Rated Capacity
The number of amperehours a cell can deliver under specific conditions (rate of discharge, end voltage, temperature); usually the manufacturer's rating.

Rated Capacity (Battery)
The number of Amp-Hours a battery can deliver under specific conditions (rate of discharge, end voltage, temperature).

Rated Lamp Life
With regard to lighting, the point in time when 50% of a statistically significant number of lamps has failed.

Rated Module Current (A)
The current output of a photovoltaic module measured at standard test conditions of 1,000 w/m2 and 25?C cell temperature.

Rated Output
The output at standard calibration.

Rated Power
Rated power of the inverter. However, some units can not produce rated power continuously.

Rated Power Output
Used by wind generator manufacturers to provide a baseline for measuring performance. Rated output may vary by manufacturer. For example, one manufacturer's 1500 watt turbine may produce that amount of power at a 30 mph windspeed, while another brand of 1500 watt turbine may not make 1500 Watts until it gets a 40 mph windspeed! So read manufacturer's ratings statements very carefully.

Rated Voltage
The maximum voltage at which an electric component can operate for extended periods without undue degradation or safety hazard.

Ratemaking Authority
A utility commission's legal authority to fix, modify, approve, or disapprove rates, as determined by the powers given the commission by a State or Federal legislature.

Residual Current Device (RCD)
A protection device which is actuated by the residual current.

Reactance
The opposition of inductance and capacitance to alternating current equal to the product of the sine of the angular phase difference between the current and voltage.

Reaction Turbine
Reaction turbines produce power from the pressure of water 'falling' on the runners after flowing through the guide vanes. Reaction turbines can operate with head as low as two feet, but require much higher flow rates than impulse turbines.

Reactive Power
The mathematical product of voltage and current consumed by reactive loads. Examples of reactive loads include capacitors and inductors. These types of loads when connected to an

ac voltage source will draw current, but since the current is 90° out of phase with the applied voltage they actually consume no real power in the ideal sense.

Reactor

A device for introducing inductive reactance into a circuit for motor starting, operating transformers in parallel, and controlling current.

Read

The process of retrieving information from a memory.

Reading

The expected output at a given input value.

Real Power

The rate at which work is performed or that energy is transferred. Electric power is commonly measured in watts or kilowatts. The term real power is often used in place of the term power alone to differentiate from reactive power. Also called active power.

Real-time clock (RTCs)

Integrated circuit that contains a timer that supplies the time of day (and often, the date). An RTC generally contains a long-life battery to allow it to keep track of the time even when there is no power applied.

Receptacle

A socket or outlet into which a plug can be inserted to make an electrical connection.

Rechargeable

Capable of being recharged; refers to secondary cells or batteries.

Recloser

A switching device that rapidly recloses a power switch after it has been opened by an overload. In reclosing the power feed to the line, the device tests the circuit to determine if the problem is still there. If not, power is not unnecessarily interrupted on the circuit.

Recovered Energy

Reused heat or energy that otherwise would be lost. For example, a combined cycle power plant recaptures some of its own waste heat and reuses it to make extra electric power.

Recovery Voltage

The voltage impressed upon

the fuse after a circuit is cleared.

Recrystallization

The formation of a new set of strain-free grains within a previously cold-worked material; normally an annealing heat treatment is necessary.

Rectification

The conversion of ac to pulsating dc.

Rectifier Type Meter

A moving coil meter used together with a bridge rectifier to measure the average value of the waveform. They are usually calibrated to read the rms value of a sinusoidal waveform by multiplication by the form factor 1.1107 during calibration.

Recycle Mark

A design of three arrows that make up a circle. This mark tells you that you can recycle the product. It can also mean that the material is made from recycled materials.

Recycling

A way to reuse materials instead of just throwing them away.

Reduced Low Voltage System

A system in which the nominal phase to phase voltage does not exceed 110 volts and the nominal phase to earth voltage does not exceed 63.5 volts.

Reduction

A chemical process that results in the acceptance of electrons by an electrode's active material.

Redundancy

The inclusion of additional assemblies and circuits (as within a UPS) with provision for automatic switchover from a failing assembly or circuit to its backup counterpart.

Reed Sensor

A reed sensor is a device built using a reed switch with additional functionality like ability to withstand higher shock, easier mounting, additional intelligent circuitry, etc.

Reed Switch

A reed switch is a passive device consisting of two reed contacts sealed inside a glass tube which operates when brought near a magnetic field.

Reel
A cylinder device used to hold wire and cable until installed. There are standard reel sizes that are used in the electrical industry that are either wood (non-returnable) or steel (returnable).

Reference range
A specific range of values of an influence quantity within which the transducer complies with the requirements concerning intrinsic errors.

Reference value
A specified single value of an influence quantity at which the transducer complies with the requirements concerning intrinsic errors.

Refererence Conditions
Conditions of use for a transducer prescribed for performance testing, or to ensure valid comparison of results of measurement.

Reflection
The return wave generated when a traveling wave reaches a load, a source, or a junction where there is a change in line impedance.

Reflector
A device used to re-direct the luminous flux from a light source by the process of reflection.

Refraction
Bending of a light beam upon passing from one medium into another.

Refractive Index
The ratio of the velocity of light in a vacuum to the velocity of light in some medium.

Refractor
A device used to re-direct the luminous flux from a light source by the process of refraction.

Refractory
A metal or ceramic that may be exposed to extremely high temperatures without deteriorating rapidly or without melting.

Refresh
The process of renewing the contents of a dynamic memory.

Regional Power Exchange
An entity established to coordinate short-term operations to maintain system stability and achieve least-cost dispatch. The dispatch pro-

vides back-up supplies, short-term excess sales, reactive power support, and spinning reserve. The pool may own, manager and/or operate the transmission lines or be an independent entity that manages the transactions between entities.

Regional Reliability Councils

Regional organizations charged with maintaining system reliability even during abnormal bulk power conditions such as outages and unexpectedly high loads.

Regional Transmission Group

A utility industry concept that the Federal Energy Regulatory Commission embraced for the certification of voluntary groups that would be responsible for transmission planning and use on a regional basis.

Register

A group of flip-flops capable of storing data.

Regulating Transformer

A transformer used to vary the voltage, or phase angle, of an output circuit. It con-

trols the output within specified limits and compensates for fluctuations of load and input voltage.

Reinforced Insulation

Single insulation system applied to live parts which provide a degree of protection against electric shock equivalent to double insulation under the conditions specified in the relevant standard. The term 'single insulation' does not imply that the insulation must be one homogeneous piece. It may comprise several layers which cannot be tested singly as supplementary or basic insulation.

Relative Magnetic Permeability

The ratio of the magnetic permeability of some medium to that of a vacuum.

Relative Permittivity

Ratio of the permittivity of the dielectric material to that of free space. The permittivity of free space is 8.854×10^{-12} F/m. The permittivity of air is very nearly equal to the permittivity of free space.

Relay

An electromagnetic device

which permits control of current in one circuit by a much smaller current in another circuit.

Relay Numerical

A protection relay which utilizes a digital signal processor to execute the protection algorithms in software.

Relays Distance

Relays used on transmission lines that use a variety of sensors and measurements to determine when an unusual condition exists at some distance, out on the transmission circuit.

Release AT (RAT)

The measured value, in AT, at which a reed contact opens. This is valid for the opening of form A, B, and E type reed contacts, and the change over from the closed normally open contact to the open normally closed contact for form C and D type contacts.

Release Time

The time interval from coil de-energization to the opening or change over of the reed contact. Where not otherwise stated, the functioning time of the reed contact in question is taken as its initial functioning time, not including contact bounce.

Reliability

Electric system reliability has two components-adequacy and security. Adequacy is the ability of the electric system to supply the aggregate electric demand and energy requirements of the consumers at all times, taking into account scheduled and unscheduled outages of system facilities. Security is the ability of the electric system to withstand sudden disturbances such as electric short circuits or unanticipated loss of system facilities.

Reliability Councils

Regional reliability councils were organized after the 1965 northeast blackout to coordinate reliability practices and avoid or minimize future outages. They are voluntary organizations of transmission-owning utilities and in some cases power cooperatives, power marketers, and non-utility generators. Membership rules vary from region to region. They are coordinated through the North American Electric Reliability Council.

Reluctance

The reluctance of a magnetic material is the ability to oppose the flow of magnetic flux. It is the constant of proportionality between the applied mmf and the flux produced. [Unit: henry^{-1}]

Remaining Relative Capacity (RRC)

The percent of the full charge that remains in a power cell.

Remanence

The residual magnetisation of a ferromagnetic substance subjected to a hysteresis cycle when the magnetising field is reduced to zero.

Remote Control

The Control of an electrical device from a remote point.

Remote Diode

A diode or diode-connected bipolar transistor used as a temperature-sensing element, often integrated onto an integrated circuit whose temperature is to be measured.

Remote Temperature

Temperature at a location other than at the die of the temperature-measuring integrated circuit.

Remote Temperature Sensor

A remotely located PN junction used as a temperature sensing device, usually located on an integrated circuit other than the one doing the measurement.

Remote Terminal Unit (RTU)

An IED used specifically for interfacing between a computer and other devices. Sometimes may include control, monitoring, or storage functions.

Renewable Energy

Electricity produced by sources that are capable of being replaced naturally and do not involve the burning of fossil fuels or the use of nuclear energy. Renewable energy is considered less harmful to the environment because it results in lower air pollution, emissions and waste than electricity from traditional sources. Some examples of renewable energy sources are solar, wind and geothermal (heat from the earth) sources and from the burning of biomass (agricultural or other wastes).

Renewable Fuels

Fuels that can be easily

made or "renewed." We can never use up renewable fuels. Types of renewable fuels are solar, wind, and hydropower energy.

Renewable Resources

Naturally, but flow-limited resources that can be replenished. They are virtually inexhaustible in duration but limited in the amount of energy that is available per unit of time. Some (such as geothermal and biomass) may be stock-limited in that stocks are depleted by use, but on a time scale of decades, or perhaps centuries, they can probably be replenished. Renewable energy resources include biomass, hydro, geothermal, solar and wind. In the future, they could also include the use of ocean thermal, wave, and tidal action technologies. Utility renewable resource applications include bulk electricity generation, on-site electricity generation, distributed electricity generation, non-grid-connected generation, and demand-reduction (energy efficiency) technologies.

Repowered Plant

This is an existing power facility that has been sub-

stantially rebuilt to extend its useful life.

Resealable Cap (Battery)

A safety vent valve which is capable of closing after each pressure release from within a cell.

Reserve Capacity

The amount of generating capacity a central power system must maintain to meet peak loads.

Reserve Generating Capacity

The amount of power that can be produced at a given point in time by generating units that are kept available in case of special need. This capacity may e used when unusually high power demand occurs, or when other generating units are off-line for maintenance, repair or refueling.

Reserve Margin

The percentage of installed capacity exceeding the expected peak demand during a specified period.

Resetting Value

The limiting value of the characteristic quantity at which the relay returns to its initial position.

Residual Current

The algebraic sum, in a multi-phase system, of all the line currents.

Residual Current Device (RCD)

Devices are designed to protect both equipment and users from fault currents between the live and earth conductors. An RCD detects the residual current between the live and neutral conductors and prevents fatal electric shocks by disconnecting the supply if the detected current exceeds a safe limit (typically 30mA).

Residual Voltage

The algebraic sum, in a multi-phase system, of all the line-to-earth voltages.

Resin

Natural resins are amorphous organic compounds which are secreted by certain plants and insects. They are usually insoluble in water but soluble in various organic solvents.

Resistance (R)

The property of a conductor, which opposes the flow of an electric current resulting in the generation of heat in the conducting material. The measure of the resistance of a given conductor is the electromotive force needed for a unit current flow. The unit of resistance is ohms.

Resistance Drop

The voltage drop in place with the current.

Resistance Thermometer

A device to measure temperature using the temperature coefficient of the material of the device (usually platinum wire).

Resistive Voltage Drop

The voltage developed across a cell by the current flow through the resistance of the cell.

Resistivity

The reciprocal of electrical conductivity, and a measure of a material's resistance to the passage of electric current.

Resistor

An aggregation of one or more units possessing the property of electrical resistance. Resistors are used in electric circuits for the purpose of operation, protection, or control.

Resolution

The smallest change that can occur in the output for a change in the input.

Resonance

A condition in which a quantity reaches maximum value. In electrical circuits, it is a condition in an RLC circuit in which the magnitude of the voltage or the current becomes a maximum or the circuit becomes purely resistive.

Resonance Frequency (in Hz)

The maximum operating frequency that a reed contact can withstand, after which it chatters, or starts sympathetic vibration.

Response Time

The time for a sensor to respond from no load to a step change in load. Usually specified as time to rise to 90% of final value, measured from onset of step input change in measured variable.

Rest Mass

The mass of a body when at rest relative to the observer.

Return Stroke

The neutralising and main stroke in a lightning strike.

Return to Zero (RZ)

A binary bitstream encoding scheme in which the signal returns to zero voltage in between the data bits. The signal has three valid levels: High, Low, and the return to zero volts after each bit.

Revenue

The total amount of money received by an organisation from sales of its products and/or services, gains from the sales or exchange of assets, interest and dividends earned on investments, and other increases in the owner's equity except those arising from capital adjustments.

Reverse Bias

Condition where the current producing capability of a PV cell is significantly less than that of other cells in its series string. This can occur when a cell is shaded, cracked, or otherwise degraded or when it is electrically poorly matched with other cells in its string.

Reverse Current

Current in a circuit of a semiconductor device due to conduction by minority carriers across PN junction.

Reverse Current Protection

Any method of preventing unwanted current flow from the battery to the photovoltaic array (usually at night).

Reversible Output Current

An output current which reverses polarity in response to a change of sign or direction of the measurand.

Rewirable Fuse

It is a semi-enclosed simple fuse which can be re-wired. It consists of a short length of wire, generally of tinned copper.

Rheostat

An adjustable resistor constructed so that its resistance may be changed without opening the circuit.

Ribbon

A thin sheet of crystalline or multicrystalline material, such as silicon, produced in a continuous process by withdrawal from a molten bath of the parent material.

Ribbon (Photovoltaic) Cells

A type of photovoltaic device made in a continuous process of pulling material from a molten bath of photovoltaic material, such as silicon, to form a thin sheet of material.

Ride through

The ability of a power conditioner to supply output power when input power is lost.

Ridge Pin

A device that allows the mounting of a pin type insulator to a pole. The ridge pin is bolted to the top of the pole and the insulator is screwed onto the threads at its top.

Ring Circuit

Each circuit commences from consumer unit (or distribution board) through an MCB (or fuse) of specific rating usually 30 A, loops into each socket outlet and returns to the same MCB (or fuse) in the consumer unit (distribution board). Looping must be done for the live conductor, neutral conductor and the protective conductor in separate rings. The ring method of connection is done only for the 13 A socket outlets, as the individual 13A plugs are separately having fuses (fuses may be usually rated at 13 A or 3 A depending on the type of load).

Ripple

The magnitude of AC fluctuation in a DC signal, after filtering. Ripple is usually expressed as a percentage of rated output.

Ripple Content of the Output

With steady-state input conditions, the peak-to-peak value of the fluctuating component of the output.

Rise Time

Usually defined as the time taken for the leading edge of a pulse or waveform to rise from 10% to 90% of its final value.

Riser Pole

A pole used to transition from overhead and underground cables.

Root-Mean-Square Value or Rms Value

The rms value of a periodic waveform is obtained by taking the square root of the mean of the squared waveform. It is also the same as the effective value of the waveform. For a.c. waveforms, unless otherwise specified, it is always the rms value that is specified.

Rolling Blackouts

A controlled and temporary interruption of electrical ser-

vice. These are necessary when a utility is unable to meet heavy peak demands because of an extreme deficiency in power supply.

Room Index K

Code number, representative of the geometry of a room, used in calculation of the utilization factor or the utilance. The room index is usually given by the formula $K = (l \times b)/h(l + b)$ where: l is the length of the room, b is the width and h is the distance of the luminaires above the working plane.

Root

The area of a blade nearest to the hub. Generally the thickest and widest part of the blade.

Rotary

A switch where rotating the actuator in a clockwise direction makes the circuit connection, and then rotating the actuator in either the same or opposite direction breaks the connection.

Rotor

(1) The blade and hub assembly of a wind generator, (2) The disc part of a vehicle disc brake, (3) The armature

of a permanent magnet alternator, which spins and contains permanent magnets.

Router

Device utilized to route data from one local-area network to another or to a phone line's long-distance line.

Routine Tests

Routine tests are conducted on each and every single equipment to ensure that each equipment meets a minimum standard of quality.

Runner

The part of a micro hydro turbine that actually attaches to the alternator drive shaft. The 'buckets' on the runner are what the water pushes against to turn the runner and generate electricity.

Running and Quick-Start Capability

The net capability of gener-

ating units that carry load or have quick-start capability. In general, quick-start capability refers to generating units that can be available for load within a 30-minute period.

Rupture

Failure that is accompanied by significant plastic deformation.

Rural Electric Cooperative or Electric Cooperative

A member-owned electric utility company serving retail electricity customers. Electric cooperatives may be engaged in the generation, wholesale purchasing, transmission, and/or distribution of electric power to serve the demands of their members on a not-for-profit basis.

R-value

Thermal resistance to the transfer or flow of heat. Insulating material is rated by its ability to resist heat flow.

Sacrificial Anode

A piece of metal buried near a structure that is to be protected from corrosion. The metal of the sacrificial anode is intended to corrode and reduce the corrosion of the protected structure.

Safety or Tamper-Resistant

A receptacle specially constructed so that access to its energized contacts is limited. Tamper-resistant receptacles are required by the National Electric Code NFPA-70 in specific pediatric care areas in health care facilities.

Sag

A voltage sag is a momentary (less than 2 seconds and more than 1 cycle) decrease in voltage outside the normal tolerance. Voltage sags are often caused by starting heavy loads, such as motors or welding equipment, and by power system faults.

Samples per Second

1. sps: Samples per second. In data conversion, an analog signal is converted to a stream of numbers, each representing the analog signal's amplitude at a moment in time. Each number is called a "sample." The number sample per second is called the sampling rate, measured in samples per second.
2. ksps: Kilosample(s) per second (thousands of samples per second)
3. Msps: Megasamples per second (millions of samples per second)

Sampling Rate

An A/D converter converts an analog signal into a stream of digital numbers, each representing the analog signal's amplitude at a moment in time. Each number is called a "sample." The number sample per second is called the sampling rate, measured in samples per second.

Saponification
The chemical process of forming a soap. More particularly a deterioration by softening of caused by the action of aqueous alkali on fatty-acid constituents.

Satellite Power System (SPS)
Concept for providing large amounts of electricity for use on the Earth from one or more satellites in geosynchronous Earth orbit. A very large array of solar cells on each satellite would provide electricity, which would be converted to microwave energy and beamed to a receiving antenna on the ground. There, it would be reconverted into electricity and distributed the same as any other centrally generated power, through a grid.

Saturation
Magnetic saturation exists when an increase of magnetization applied to a reed contact does not increase the magnetic flux through.

Savings Fraction
The percentage of consumption from using the old technology that can be saved by replacing it with the new, more efficient demand-side management technology. For example, if a 60-watt incandescent lamp were replaced with a 15-watt compact fluorescent lamp, the savings fraction would be 75 percent because the compact fluorescent lamp uses only 25 percent of the energy used by the incandescent lamp.

Scaling Resistor
A resistor added to an output circuit of measurement equipment to provide a scaled voltage output. The output is not a "true" voltage output and may be susceptible to loading errors.

Scheduled Outage
The shutdown of a generating unit, transmission line, or other facility, for inspection or maintenance, in accordance with an advance schedule.

Scheduling Coordinators
Entities certified by the Federal Energy Regulatory Commission that act as a go-between with the Independent System Operator on

behalf of generators, supply aggregators (wholesale marketers), retailers, and customers to schedule the distribution of electricity.

Schematic Diagram
A diagram which shows, by means of graphic symbols, the electrical connections and functions of a curcuit.

Schmitt Trigger
A circuit devised of an op-amp configured as a comparator and given a positive feedback. The schmitt trigger employs hysteresis in order to create a switching voltage which is less susceptible to noise than a straight switch.

Schottky Barrier
A cell barrier established as the interface between a semiconductor, such as silicon, and a sheet of metal.

Scribing
The cutting of a grid pattern of grooves in a semiconductor material, generally for the purpose of making interconnections.

Scrubbers
Equipment designed to reduce sulfur emissions from coal-fired generating plants.

Seal
The structural part of a galvanic cell that restricts the escape of solvent or electrolyte from the cell and limits the ingress of air into the cell (the air may dry out the electrolyte or interfere with the chemical reactions).

Sealed Battery
A battery with a captive electrolyte and a resealing vent cap, also called a valve-regulated battery. Electrolyte cannot be added.

Seasonal Depth of Discharge
An adjustment factor used in some system sizing procedures which "allows" the battery to be gradually discharged over a 30-90 day period of poor solar insolation. This factor results in a slightly smaller photovoltaic array.

Seasonal Energy-Efficiency Ratio (SEER)
The ratio of the total seasonal cooling requirement (measured in Btu) to the total seasonal Wh of energy used, expressed in terms of Btu/Wh. (The SEER rating equals 3.413 times the seasonal COP.)

Second Harmonic Distortion
Ratio of second-order har-

monic to the input signal (carrier). Often measured as dBc.

Second(s)

The second is the SI unit of time. It is a fundamental unit. It is defined as the duration of 9 192 631 770 periods of the radiation corresponding to the transition between the two hyperfine levels of the ground state of the caesium 133 atom [1967]

Secondary

That winding of a transformer which receives its energy by electromagnetic induction from the primary. Also, frequently referred to as the output winding. A transformer may have one or more secondaries.

Secondary Battery

A battery made up of secondary cells.

Secure Hash Algorithm (SHA)

A message digest algorithm developed by the NSA for use in the Digital Signature standard, FIPS number 186 from NIST. SHA is an improved variant of MD4 producing a 160-bit hash. SHA is one of two message digest algorithms available in IPSEC.

Secure Hash Standard

This standard specifies a Secure Hash Algorithm, SHA-1, for computing a condensed representation of a message or a data file.

Securitization

The act of pledging assets to a creditor through a note, lien or bond. This is a mechanism to allow a utility to recover stranded costs up front in a single lump sum payment. Under a securitization scheme, the legislature or utility commission orders customers to pay a surcharge as part of their electric bill. That surcharge must be paid within the utility's original service territory, regardless of who supplies the electricity to customers.

Securitize

The aggregation of contracts for the purchase of the power output from various energy projects into one pool which then offers shares for sale in the investment market. This strategy diversifies project risks from what they would be if each project were financed individually, thereby reducing the cost of financing. Fannie

Mae performs such a function in the home mortgage market.

Selector Switch

A multi-position switch which can be set to the desired mode of operation.

Self-discharge

The tendency of all electrochemical cells to lose energy. Self-discharge represents energy lost to internal chemical reactions within the cell.

Self-Generation

A generation facility dedicated to serving a particular retail consumer, usually located on the consumer's premises. The facility may either be owned directly by the retail consumer or owned by a third party with a contractual arrangement to provide electricity to meet some or all of the consumer's load.

Self-Inductance

The ability of a circuit or coil to induce a voltage within itself.

Self-Induction

The process by which a changing current induces a voltage into the conductor or coil carrying the current.

Sellback

When an alternative energy system is connected to the grid, and excess power is sold back to the local utility.

Semiconductors

Semiconductors are the materials that are neither good conductors nor good insulators. Thy contain four valence electrons and are characterized by the fact that as they are heated, their resistance decreases. Heat has the opposite effect on conductors, whose resistance increases with an increase of temperature. Semiconductors have become extremely important in the electrical industry since the invention of the transistor in 1947. All solid state devices such as diodes, transistors, and integrated circuits are made from combinations of semiconductors materials. The two most common materials used in the production of electronic components are silicon and germanium. Of the two, silicon is used more often because of its ability to withstand heat. Before and pure semiconductor can be used to construct electronic device, it must be mixed or "doped" with an impurity.

Semi-cutoff (Lighting)

Luminaire light distribution is classified as semi-cutoff when the candlepower per 1000 lamp lumens does not numerically exceed 50 (5.0%) at an angle of 90 degrees above nadir (horizontal), and 200 (20%) at a vertical angle of 80 degrees above nadir. This applies to any lateral angle around the luminaire.

Sense Resistor

A resistor placed in a current path to allow the current to be measured. The voltage across the sense resistor is proportional to the current that is being measured and an amplifier produces a voltage or current that drives the measurement.

Sensitivity

Ratio of the output signal or response of the instrument to a change of input or measured variable.

Separated Extra Low Voltage (SELV)

An extra low voltage system which is electrically sepa-rated from Earth and from other systems in such a way that a single fault cannot give rise to the risk of electric shock.

Separately Derived System

A premises wiring system whose power is derived from a battery, a solar photovoltaic system, or from a generator, transformer, or converter windings, and that has no direct electrical connection, including solidly connected grounded circuit conductor, to supply conductors originating in another system.

Separator

The permeable membrane that allows the passage of ions, but prevents electrical contact between the anode and the cathode.

Serial Interface

A serial interface (as distinguished from a parallel interface) is one in which data is sent in a single stream of bits, usually on a single wire-plus-ground, wire-pair, or single wireless channel (or two sets, one for each direction). Examples include USB, RS-232, I^2C, and 1-Wire.

Series Circuit

A circuit supplying energy to a number of loads connected in series, that is, the same current passes through each load in completing its path to the source of supply.

Series Controller

A charge controller that interrupts the charging current by open-circuiting the photovoltaic (PV) array. The control element is in series with the PV array and battery.

Series Gap

Internal gap(s) between spaced electrodes in series with the valve elements across which all or part of the impressed arrester terminal voltage appears.

Series Motor

D.C. motor with a series connected field.

Series Parallel Circuit

An electric current containing groups of parallel connected receptive devices, the groups being arranged in the circuit in series; a series multiple circuit.

Series Regulator

Type of battery charge regulator where the charging current is controlled by a switch connected in series with the PV module or array.

Series Resistance

Parasitic resistance to current flow in a cell due to mechanisms such as resistance from the bulk of the semiconductor material, metallic contacts, and interconnections.

Series Resonance

A resonance condition usually occurring in series RLC circuits, where the current becomes a maximum for a given voltage.

Series/Multiple

A winding of two similar coils that can be connected for series operation or multiple (parallel) operation.

Service

The conductors and equipment for delivering electric energy from the serving utility to the wiring system of the premises served.

Service Cable

Service conductors made up in the form of a cable.

Service Conductors

The supply conductors that

extend from the street main or transformers to the service equipment of the premises being supplied

Service Drop

The lines running to a customer's house. Usually a service drop is made up of two 120 volt lines and a neutral line, from which the customer can obtain either 120 or 240 volts of power. When these lines are insulated and twisted together, the installation is called triplex cable.

Service Entrance

The point where the electrical service enters the house, becoming your responsibility instead of the electric company's.

Service Entrance Cable

The conductors that connect the service conductors (drop or lateral) to the service equipment of the building.

Seven Segment Display

A display consisting of 7 linear segments arranged in such a way as to be able to produce all the 10 digits from 0 to 9 by proper excitation.

Shaft

The rotating part in the cen-

ter of a wind generator or motor that transfers power.

Shallow Cycling

Charge and discharge cycles which do not allow the battery to approach it's cutoff voltage. Shallow cycling of NiCd cells lead to "memory effect". Shallow cycling is not detrimental to NiMH cells and it is the most beneficial for lead acid batteries.

Shallow-Cycle Battery

A battery with small plates that cannot withstand many discharges to a low state-of-charge.

Sheath

The covering around the insulation of a cable.

Sheet Lightning

Lightning in diffused or sheet form due to reflection and diffusion by the clouds and sky. A single bolt within a cloud or between clouds can illuminate an entire cloud from the inside. This is called sheet lightning.

Shelf Life

For a dry cell, the period of time (measured from date of manufacture), at a storage temperature of 21 degrees C

(69 degrees F), after which the cell retains a specified percentage (usually 90%) of its original energy content.

Shellac

A yellowish natural resin secreted by the lac insect which is parasitic on certain trees.

Shield

Device surrounding that portion of a connector that is used for attaching wires or cables to shield against electromagnetic interference, and/or protect connector wires or cable from mechanical damage.

Shift Register

Two or more bistable elements (flip-flops) connected in series. With each tick of the clock, the output of stage n is shifted to stage n+1. Applications include clock or signal delays, delay lines, linear-feedback shift registers.

Shock Current

A current passing through the body of a person or livestock such as to cause electric shock and having characteristics likely to cause dangerous effects.

Shock Hazard

A dangerous electrical condition associated with the possible release of energy caused by contact or approach to energized parts.

Shock Sensor

An acceleration sensor, generally a piezoelectric type, that can measure high acceleration but cannot measure static g forces.

Short Circuit

1. A load that occurs when at ungrounded conductor comes into contact with another conductor or grounded object. 2. An abnorman connection of relatively low impedance, whether made intentionally or by accident, between two points of different potential.

Short Circuit Current

An overcurrent resulting from a fault of negligible impedance between live conductors having a difference in potential under normal operating conditions.

Short Distribution (Lighting)

A luminary is classified as having a short light distribution when its max candlepower point falls between

1.0MH

2.25MH TRL. The maximum luminaire spacing-to-mounting height ratio is generally 4.5 or less.

Short for Compressor/Decompressor (CODEC)

A CODEC is any technology for compressing and decompressing data. Codecs can be implemented in software, hardware, or a combination of both.

Short Ton

A unit of weight equal to 2,000 pounds.

Short-Circuit Current

That current delivered when a cell is short-circuited (i.e., the positive and negative terminals are directly connected with a low-resistance conductor).

Shotgun Stick

A specialized hot stick that allows the capture of certain types of clamps and devices in its hook. It is also called a "Grip All" stick.

Shunt

A component connected in parallel. A current shunt is a device for altering the amount of electric current flowing through a piece of apparatus, such as a galvanometer.

Shunt Controller

A charge controller that redirects or shunts the charging current away from the battery. The controller requires a large heat sink to dissipate the current from the short-circuited photovoltaic array. Most shunt controllers are for smaller systems producing 30 amperes or less.

Shunt Motor

D.C. motor with a shunt connected field.

Shunt Regulator

Type of a battery charge regulator where the charging current is controlled by a switch or transistor connected in parallel with the PV panel. Overcharging of the battery is prevented by shorting the PV output. Shunt regulators are common in PV systems as they are relatively cheap to build and simple to design. Series regulators usually have better control and charge characteristics.

Shutdown

A feature of many Maxim/

Dallas ICs, typically controlled via a logic-level input, which dramatically reduces power consumption when the device is not in use.

Side Band

The band of frequencies lying on either side of a modulated carrier wave.

Sidewalk (lighting)

Paved or otherwise improved areas for pedestrian use, located within public street rights-of-way also containing roadways for vehicular traffic.

Sidewall Pressure

The force exerted on a cable as it is dragged around a bend. The longer the pull and the tighter the bend radius, the higher the sidewall pressure will become. High sidewall pressure damages cable. There is a higher chance of destroying cable by high sidewall pressure than by high tensile tension. Electrical cable manufacturers' specifications limit sidewall pressure to a range of 300
1500 lbs/ft, depending on cable type. There are three ways to reduce sidewall pressure while pulling cable with bends: shorten the pull, enlarge the bend radius, or use a high performance lubricant. Polywater's "Pull Planner 2000" software package automatically calculates sidewall pressure for each segment of a cable pull (a "segment" contains one straight section and one bend). This allows a designer to play "what if," easily changing pull parameters and instantly seeing the affect on sidewall pressure. For a copy of American Polywater's Pull Planner 2000, contact Young & Company.

Siemens (S)

SI unit of electric conductance. One siemens is equal to the conductance between two points of a conductor having a resistance of 1 W. siemens is the reciprocal of the ohm.

Siemens Process

A commercial method of making purified silicon.

Sigma Delta

An ADC architecture consisting of a 1-bit ADC and filtering circuitry which over-samples the input signal and performs noise-shaping to achieve a high-resolution digital output. The architecture is relatively inexpensive compared to other ADC architectures.

Signal

A visual, audible, electrical or other indication used to convey information.

Signal Conditioner

A device placed between a signal source and a readout instrument to change the signal. Examples are attenuators, preamplifiers, charge amplifiers, and sophisticated level-translating deivices that can compensate for non-linearities in the sensor or amplifier.

Signal-Invalid O/P

Signal invalid output. Indicates when all RS-232 signals to the IC are in the invalid range.

Signal-to-noise and distortion ratio (SINAD)

The RMS value of the sine wave f(IN) (input sine wave for an ADC, reconstructed output sine wave for a DAC) to the RMS value of the converter noise from DC to the Nyquist frequency, includ-

ing harmonic content. It is typically expressed in dB (decibels).

Signal-To-Noise Ratio

Signal-to-Noise Ratio, the ratio of the amplitude of the desired signal to the amplitude of noise signals at a given point in time. The larger the number, the better. Usually expressed in dB.

Silicon (Si)

A chemical element, atomic number 14, semimetallic in nature, dark gray, an excellent semiconductor material. A common constituent of sand and quartz (as the oxide). Crystallizes in face-centered cubic lattice like a diamond. The most common semiconductor material used in making photovoltaic devices.

Silicon Timed Circuit (STC)

1. A circuit that produces a delayed version of the input signal. Also known as a delay line.
2. System Timing and Control: Clock generation and distribution systems and components. May include the means for clock control such as spread-spectrum clock generation for EMI

reduction, skew rate control, rate dividers, rate control, width, delay, and phase adjustment.

Simplex Communications System

A communications system in which data can only travel in one direction.

Simultaneously Accessible Parts

Conductors or conductive parts which can be touched simultaneously by a person or, in locations specifically intended for them, by livestock. Simultaneously accessible parts may be: live parts, exposed conductive parts, extraneous conductive parts, protective conductors or earth electrodes.

Sine Wave

A sinusoidal periodic oscillation. The fundamental waveform from which other waveforms may be generated by combinations of various group of harmonics. The voltage and current waveforms produced from the power company generators (alternators) are basic sine waves.

Sine Wave Inverter

An inverter that produces utility-quality, sine wave power forms.

Single

A receptacle that accepts only one plug.

Single Chip Transceivers (SCT)

A single IC that includes data communication transmitter and receiver functions.

Single Element Transducer

A transducer having one measuring element.

Single Ended Primary Inductor Converter (SEPIC)

A DC-DC converter topology that acts both as a boost and a buck converter (that is, will step up or down, depending on the input voltage).

Single Phase Events

An electrical disturbance that affects only one phase (electricity is generated in three phases) of a two- or three-phase system.

Single Phase Line

This carriers electrical loads capable of serving the needs of residential customers, small commercial customers, and streetlights. It carrier a relatively light load as compared to heavy duty three phrase constructs.

Single Sweep

The ability of an oscilloscope to display just one window of time, thus preventing unwanted multiple displays. Necessary in the display of transient waveforms.

Single-Crystal Material

A material that is composed of a single crystal or a few large crystals.

Single-Crystal Silicon

Material with a single crystalline formation. Many photovoltaic cells are made from single-crystal silicon.

Single-Point Ground

The practice of tying the power neutral ground and safety ground together at the same point, thus avoiding a differential ground potential between points in a system.

Single-pole Switch

A standard light switch with off and on positions for controlling flow of current to one or more devices.

Single-Pole, Double-Throw (SPDT)

A switch that makes or breaks the connection of a single conductor with either of two other single conductors. This switch has 3 terminal screws, and is commonly used in pairs and called a "Three-Way" switch.

Single-Shot Reclosing

An auto-reclose sequence that provides only one reclosing operation, lockout of the CB occurring if it subsequently trips.

Single-Stage Controller

A charge controller that redirects all charging current as the battery nears full state-of-charge.

Sintered Plate (Battery)

The plate of a alkaline cell, the support of which is made of sintered metal powder, and into which the active material is introduced.

Sinusoidal

The graphical plot of the output of an alternator.

Site Acceptance Test (SAT)

Validation procedures for equipment executed with the customer on site.

Skin Effect

The tendency of current to stick to the outer layers of a conductor due to the presence of internal flux. The skin effect is more prominent at higher frequencies.

Slag

A residue produced by the combustion of coal. This heat-fused material accumulates on the sides and bottom of a boiler and is removed periodically and disposed of according to environmental regulations.

Slamming

The unauthorized switching of a customer's account to another utility or AES without the customer's consent.

Slide

A switch with a slide-action actuator for making or breaking circuit contact. Dimmer switches and fan speed controls are also available with slide-action mechanisms for lighting and fan speed control

Slimline Single-Pin

A fluorescent lampholder with a single contact designed for Slimline fluorescent lamps such as the T-12 (11/2" dia.), T-8 (1" dia.), and

the smaller version T-6 (3/4" dia.).

Slip

The difference between synchronous and operating speeds, compared to synchronous speed, expressed as a percentage. Also the difference between synchronous and operating speeds, expressed in rpm.

Slip Rings

The rotating contacts which are connected to the loops of a generator.

Slope

The ratio of change in the vertical quantity (Y) to the change in the horizontal quantity (X).

mall Computer System Interface (SCSI)

Small Computer System Interface (pronounced "scuzzy"), an interface standard for connecting peripheral devices to computers. Hardware components for implementing a SCSI interface include connector ports on computers and cables for connecting peripheral devices to the computer. SCSI is gradually being supplanted by the newer USB and IEEE-1341 standards.

Small Paddle Inductive

Inductive coupling of power from the charger to the vehicle was put on various production vehicles in two generations: first with a large paddle and then later with a small paddle. Large paddle cars can use small paddle chargers with an adapter, but the converse is not true, so SPI chargers are now much preferable to LPI.

Smart Battery

A battery with internal circuitry that provides level of charge status to the host system.

Smart Signal Conditioner

Signal conditioner that is programmable or has a flexible architecture to allow it to accomplish sophisticated signal transformations and corrections.

Smoke and Carbon Dioxide Detectors

Wall and ceiling mounted sensors located throughout the home used to alert occupants of deadly gasses and smoke inside the home.

Snap-In

An incandescent or compact fluorescent lampholder with factory-assembled spring clips that securely snap into a panel cutout without requiring additional fasteners.

Snubber

A device which suppresses voltage transients.

Socket

A connecting place or junction for electric wires, plugs and light bulbs.

Socket Outlet

A device, provided with female contacts, which is intended to be installed with the fixed wiring, and intended to receive a plug. A luminaire track system is not regarded as a socket outlet system.

Soft Start

A feature in some switching power supplies that limits the startup inrush current at initial startup.

Solar Cell

An device which converts energy from the sun directly into electrical energy.

Solar Constant

The average amount of solar radiation that reaches the earth's upper atmosphere on a surface perpendicular to the sun's rays; equal to 1353 Watts per square meter or 492 Btu per square foot.

Solar Cooling

The use of solar thermal energy or solar electricity to power a cooling appliance. Photovoltaic systems can power evaporative coolers ("swamp" coolers), heat-pumps, and air conditioners.

Solar Energy

Electromagnetic energy transmitted from the sun (solar radiation). The amount that reaches the earth is equal to one billionth of total solar energy generated, or the equivalent of about 420 trillion kilowatt-hours.

Solar Noon

That moment of the day that divides the daylight hours for that day exactly in half. To determine solar noon, calculate the length of the day from the time of sunset and sunrise and divide by two. Solar noon may be quite a bit different from 'clock' noon.

Solar Panel

A group of photovoltaic modules mechanically mounted on a single frame.

Solar Power

Energy from the sun's radiation converted into heat or electricity.

Solar Power Terminology

Terms that apply specifically to solar power systems, particularly photovoltaic systems.

Solar Resource

The amount of solar insolation a site receives, usually measured in kWh/m2/day, which is equivalent to the number of peak sun hours.

Solar Spectrum

The total distribution of electromagnetic radiation emanating from the sun. The different regions of the solar spectrum are described by their wavelength range. The visible region extends from about 390 to 780 nanometers (a nanometer is one billionth of one meter). About 99 percent of solar radiation is contained in a wavelength region from 300 nm (ultraviolet) to 3,000 nm (near-infrared). The combined radiation in the wavelength region from 280 nm to 4,000 nm is called the broadband, or total, solar radiation.

Solar Thermal Electric

A process that generates electricity by converting incoming solar radiation to thermal energy.

Solar-Grade Silicon

Intermediate-grade silicon used in the manufacture of solar cells. Less expensive than electronic-grade silicon.

Solenoid

A spiral of conducting wire, would cylindrically so that when an electric current passes through it, its turns are nearly equivalent to a succession of parallel circuits, and it acquires magnetic properties similar to those of a bar magnet.

Solid Angle

The angle subtended at the center of a sphere by an area on its surface numerically equal to the square of the radius. [Unit steradian, sr].

Solid State Relay

A completely electronic switching device with no moving parts or contacts.

Source Energy

All the energy used in delivering energy to a site, including power generation and transmission and distribution losses, to perform a specific function, such as space conditioning, lighting, or water heating. Approximately three watts (or 10.239 Btus) of energy is consumed to deliver one watt of usable electricity.

Southeastern Electric Reliability Council (SERC)

One of the ten regional reliability councils that make up the North American Electric Reliability Council (NERC).

Southwest Power Pool (SPP)

One of the ten regional reliability councils that make up the North American Electric Reliability Council (NERC).

Space Charge

In an electron tube, a cloud of free electrons surrounding the emitter.

Spacing-to-Mounting Height Ratio

Ratio specification used to insure that fixtures are adequately spaded, thus preventing "hotspots"

Span

1) Refers to the distance between two poles of a transmission or distribution line.

2) The algebraic difference between the upper and lower values of a range.

Spark
A discharge of electricity across a gap between two electrodes. The discharge is accompanied by heat and incandescence. Distinguish between spark and arc.

Spark Test
A high-voltage test performed on certain types of conductor during manufacture to ensure the insulation is free from defects.

Specialized Mobile Radio (SMR)
Indicates the 896MHz to 901MHz band (800MHz band), which uses two paired 25kHz channels, and the 935MHz to 940MHz band (900 MHz band), which uses two paired 12.5kHz channels. Ten 20-channel blocks have been allocated in these frequency bands by the FCC. 900MHz SMR is primarily used for radio dispatch, paging, and wireless data communications.

Specific Conjuctive Test
A conjunctive test using specific values of each of the parameters.

Specific Heat
The quantity of heat required to raise the temperature of unit mass through 1°C.

Specific-Gravity (Battery)
The weight of the electrolyte compared to the weight of an equal volume of pure water. It is used to measure the strength or percentage of sulfuric acid in th electrolyte.

Speed of Light
Speed of light in vacuum (c_o)
= $2.997\ 925 \times 108$ m s^{-1}

Sphere Gap
A gap between two spherical electrodes. The sphere gap method of measuring high voltage is the most reliable and is used as the standard for calibration purposes.

Spheroid
Solid figure generated by an ellipse rotating about one of its axes.

Spike (or Impulse, Switching Surge, Lightning Surge)
These terms refer to a voltage increase of very short duration (microsecond to millisecond). Spikes can range in amplitude from 200

V to 6,000 V and are caused by lightning, switching of heavy loads, and short circuits or power system faults.

Spike Suppressor

A device that provides protection against short duration (microsecond to millisecond) voltage increases known as spikes, impulses, transients, or high-frequency surges.

Spill Light

Unwanted light directed onto a neighboring property. Also referred to Light Trespass.

Spinning Reserve

That reserve generating capacity running at a zero load and synchronized to the electric system.

Split Phase

A split phase electric distribution system is a 3-wire single-phase distribution system, commonly used in North America for single-family residential and light commercial (up to about 100 kVA) applications. It is the AC equivalent of the former Edison direct current distribution system. Like that system, it has the advantage of saving the weight of conductors for the installation. Since there are two live conductors in the system, it is sometimes incorrectly referred to as "two phase".

Split Receptacle

A receptacle in which each of the two outlets is wired on a different circuit or in which one outlet is always live and the other is switched. Also called *split-wired*.

Split-Circuit

A duplex receptacle that allows each receptacle to be wired to separate circuits. Most duplex receptacles provide break-off tabs that allow them to be converted into split-circuit receptacles.

Split-Spectrum Cell

A compound photovoltaic device in which sunlight is first divided into spectral regions by optical means. Each region is then directed to a different photovoltaic cell optimized for converting that portion of the spectrum into electricity. Such a device achieves significantly greater overall conversion of incident sunlight into electricity.

Split-the-savings

The basis for settling economy-energy transactions between utilities. The added cost of the supplier are subtracted from the avoided costs of the buyer, and the difference is evenly divided.

Spotlight

A (small) projector giving concentrated light of usually not more than 20° divergence.

Spread Spectrum

A wireless communications technology that scatters data transmissions across the available frequency band in pseudorandom pattern. Spreading the data across the frequency spectrum greatly increases the bandwidth which in turn can reduce noise and provide privacy.

Spring Winding Time

For spring-closed CB's, the time for the spring to be fully charged after a closing operation.

Sprites

Massive but weak luminous flashes that appear directly above an active thunderstorm system and are coincident with cloud-to-ground or intra-cloud lightning strokes. Sprites are immense. They can shoot up from the top of a 8 km thundercloud to heights of 40 km or more.

Spur

A branch from a ring final circuit.

Spurious-Free

Unwanted frequencies are not present.

Sputtering

A physical vapor deposition process where high-energy ions are used to bombard elemental sources of semiconductor material, which eject vapors of atoms that are then deposited in thin layers on a substrate.

Square Wave

A periodic wave which alternates between two fixed amplitudes for equal lengths of time, with the time of transition between the amplitudes being negligibly small.

Square Wave Inverter

The inverter consists of a dc source, four switches, and the load. The switches are power semiconductors that can carry a large current and withstand a high voltage rating. The switches are turned on and off at a correct sequence, at a certain frequency. The square wave inverter is the simplest and the least expensive to purchase, but it produces the lowest quality of power.

Squelch

A circuit which mutes the signal when it is below a certain level. Typically used to quiet the signal when only noise is present.

Squirrel Cage Rotor

This type of rotor has the simplest and most rugged construction and is almost indestructible. The rotor consists of a cylindrical core with parallel slots for carrying the rotor conductors which are not wires but heavy bars of copper, aluminium or alloys. The rotor bars are permanently short-circuited at the ends to form the winding or cage. About 90% of induction motors are squirrel cage type.

Stability

The property of a system or element by virtue of which its output will ultimately attain a steady state. The sta-bility of a power system is its ability to develop restoring forces equal to or greater than the disturbing forces so as to maintain a state of equilibrium.

Staebler-Wronski Effect

The tendency of the sunlight to electricity conversion efficiency of amorphous silicon photovoltaic devices to degrade (drop) upon initial exposure to light.

Stand-Alone System

An autonomous or hybrid photovoltaic system not connected to a grid. May or may not have storage, but most stand-alone systems require batteries or some other form of storage.

Standard calibration

The nominal point at which a measurement device is adjusted.

Standard Cell

A specially prepared primary cell which is characterised by a highly constant emf over long periods of time.

Standby Current

This is the amount of current (power) used by the inverter

when no load is active (lost power). The efficiency of the inverter is lowest when the load demand is low.

Standby Facility

A facility that supports a system and generally running under no load.

Standby Power Supply

The power supply that is available to furnish electric power when the normal power supply is not available.

Standby Service

A customer or AES may at their option contract with a utility or a competitive supplier to provide electric generation service in the event that the primary source of supply is disabled or becomes unavailable for any reason. This is sometimes called "standby generation service." The MPSC establishes the rates regulated electric suppliers can charge for providing standby service. Competitive suppliers may provide standby service under market-based rates.

Standby Uninterruptible Power Supply (UPS)

A type of uninterruptible power supply (UPS) that re-

mains in standby mode offering little or no power conditioning unless the voltage exceeds a set limit. If the voltage limit is exceeded, the UPS can support the load until the voltage returns to within the limits or the battery is drained.

Stand-Off Mounting

Technique for mounting a photovoltaic array on a sloped roof, which involves mounting the modules a short distance above the pitched roof and tilting them to the optimum angle.

Star Connection

A method of connecting three elements of a three-phase electrical system at a common node, and with the three phases being taken from the remaining nodes of the elements. This is also known as "wye" connection.

Star Ground

A pcb layout technique in which all components connect to ground at a single point. The traces make in a "star" pattern, emanating from the central ground.

Star Point

A point from which all

traces leave in a "star" pattern in pcb layout.

Starter

A device used in conjuction with a ballast for the purpose of starting an electric discharge lamp.

Starting Current

Current required by the ballast during initial arc tube ignition. Current changes as lamp reaches normal operating light level.

Starting Relay

A unit relay which responds to abnormal conditions and initiates the operation of other elements of the protection system.

Starting Torque

The torque produced by a motor at rest when power is applied. For an AC machine, this is the locked-rotor torque.

Starting-Lighting-Ignition (SLI) Battery

A battery designed to start internal combustion engines and to power the electrical systems in automobiles when the engine is not running. SLI batteries can be used in emergency lighting situations.

Start-Up

The windspeed at which a wind turbine rotor starts to rotate. It does not necessarily produce any power until it reaches cut-in speed.

Starved Cell (Battery)

A cell containing little or no free fluid electrolyte solution. This enables gasses to reach electrode surfaces readily, and permits relitive high rates of recombination.

Starved Electrolyte Cell

A battery containing little or no free fluid electrolyte.

State of Charge (Battery)

The available amp-hours in a battery at any point of time. State of Charge is determined by the amount of sulfuric acid remaining in the electrolyte at the time of testing or by the stabilized open circuit voltage.

Static Contact Resistance (CR)

The electrical resistance of closed reed contacts, as measured terminal to terminal, at their associated terminals.

Static Electricity

Created when electrons "jump" from one atom to

another. You can create static electricity by rubbing certain things together, such as a brush and your hair. Lightning is also an example of static electricity.

Static Pressure

Pressure produced by an unmoving column of water. (also: static head) There are no friction/head losses when water is not moving, so static pressure is determined only by the vertical height of the water column. The static pressure on a 10ft. tall vertical pipe full of water would be the same as a 1000ft. long pipeline with 10ft. of head over its entire distance.

Static Relay

An electrical relay in which the designed response is developed by electronic, magnetic, optical or other components without mechanical motion. Excludes relays using digital technology.

Static Var Compensator (STATCOM)

A particular type of Static Var Compensator, in which Power Electronic Devices such as GTO's are used to generate the reactive power

required, rather than capacitors and inductors.

Static Wire

A wire placed above the phase wires of a distribution of transmission circuit to protect against lightning. It is normally galvanized or aluminized steel.

Stationary

With wind generator towers, a tower that does not tilt up and down. The tower must be climbed or accessed with a crane to install or service equipment at the top.

Stationary Battery

A secondary battery designed for use in a fixed location.

Stationary Equipment

Electrical equipment which is either fixed, or equipment having a mass exceeding 18 kg and not provided with a carrying handle.

Statistical Impulse Voltage

This is the switching or lightning overvoltage applied to equipment as a result of an event of one specific type on the system (line energising, reclosing, fault occurrence, lightning discharge, etc), the peak value of which has a 2% probability of being exceeded.

Statistical Impulse Withstand Voltage

This is the peak value of a switching or lightning impulse test voltage at which insulation exhibits, under the specified conditions, a 90% probability of withstand. In practice, there is no 100% probability of withstand voltage. Thus the value chosen is that which has a 10% probability of breakdown.

Stator

The stationary member of a machine in the form of an hollow cylinder inside which the rotor will be placed with a narrow intervening air gap.

Steady Current

An electric current of constant amperage.

Steady State Response

Behaviour of a circuit after a long time when steady conditions have been reached after an external excitation.

Steam Electric Plant (Conventional)

A plant in which the prime mover is a steam turbine. The steam used to drive the turbine is produced in a boiler where fossil fuels are burned.

Steam Plant (Conventional)

A plant in which the prime mover is a steam turbine. The steam used to drive the turbine is produced in a boiler where fossil fuels are burned.

Steam-Electric Plant (Conventional)

A plant in which the prime mover is a steam turbine. The steam used to drive the turbine is produced in a boiler where fossil fuels are burned.

Step Leaders

Thin, luminous feelers, caused by electrical breakdown in a cloud, that move in short bursts, or steps, and precede lightning strikes. Lightning begins in the negatively charged region at the base of a cloud. Here, thin, barely luminous feelers called step leaders zigzag through the cloud and can travel to the earth.

Step Response

Behaviour of a circuit when the excitation is the unit step

function. The excitation function may be a voltage or a current.

Step Voltage

The difference in surface potential experienced by a person bridging a distance of 1 m with his feet without contacting any other grounded structure.

Step Waveform

A waveform which has one level (usually zero) prior to zero time and another level after time zero.

Step-down

This refers to a transformer that has fewer turns of wire in the secondary than in the primary, which causes a decrease or step-down of the voltage.

Step-up

This refers to a transformer that has more turns of wire in the secondary than in the primary, which causes an increase or step-up of the voltage.

Step-Up DC-DC

A switch-mode voltage regulator in which output voltage is higher than its input voltage.

Steradian (sr)

The steradian is SI unit of the solid angle. It is a supplementary unit. It is defined as the solid angle that, having its vertex in the centre of a sphere, cuts off an area of the surface of the sphere equal to that of a square with sides of length equal to the radius of the sphere.

Stinger

Slang for the wire connecting a fused cutout or switch to a transformer bushing.

Stocks

A supply of fuel accumulated for future use. This includes coal and fuel oil stocks at the plant site, in coal cars, tanks, or barges at the plant site, or at separate storage sites.

Storage Battery

An assembly of identical cells in which the electrochemical action is reversible so that the battery may be recharged by passing a current through the cells in the opposite direction to that of discharge. While many non-storage batteries have a reversible process, only those that are economically rechargeable are classified as storage batteries. Synonym: Accumulator; Secondary Battery.

Storage Cell

An electrolytic cell for the generation of electric energy in which the cell after being discharged may be restored to a charged condition by an electric current flowing in a direction opposite the flow of current when the cell discharges. Synonym: Secondary Cell.

Storage Conditions

The conditions defined by means of ranges of the influence quantities, such as temperature, or any special conditions, within which the transducer may be stored (non-operating) without damage.

Storm Windows

An extra pane of glass or plastic added to a window to reduce air infiltration and boost the insulation value of a window. If you are considering adding storm windows, you should compare the costs to installing new energy-efficient windows.

Straight Blade

A non-locking connector into which mating plugs are inserted at a right angle to the plane of the connector face.

Strand

One of the wires that made up a stranded conductor.

Strategic Conservation

Strategic conservation results from load reductions occurring in all or nearly all time periods. This strategy can be induced by price of electricity, energy-efficient equipment, or decreasing usage of equipment.

Strategic Load Growth

A form of load building designed to increase efficiency in a power system. This load shape objective can be induced by the price of electricity and by the switching of fuel technologies (from gas to electric).

Stratification

A condition that occurs when the acid concentration varies from top to bottom in the battery electrolyte. Periodic, controlled charging at voltages that produce gassing will mix the electrolyte.

Stray Coupling

Capacitive coupling that

may occur between adjacent arms, sources, detector, leads etc.

Streamer
A ribbon like discharge.

Streamer Mechanism
The development of a spark discharge directly from a single avalanche.

String
A number of photovoltaic modules or panels interconnected electrically in series to produce the operating voltage required by the load.

Stringing Block
A sheave used to support and allow movement of a cable that is being installed. These are normally used overhead but there are also specialized designs used at the entrance to a conduit system. Stringing blocks are manufactured by Bethea.

Strobe
A pulse used for timing and synchronization.

Stroboscope
Any device used to study, measure, balance, or otherwise alter the motion of a moving, rotating, or vibrat-

ing body by making it appear to slow down or stop with the use of pulsed bursts of light or by viewing it through intermittent openings in a revolving disk.

Subbituminous Coal
A coal whose properties range from those of lignite to those of bituminous coal and are used primarily as fuel for steam-electric power generation. It may be dull, dark brown to black, soft and crumbly at the lower end of the range, to bright, jet black, hard, and relatively strong at the upper end. Subbituminous coal contains 20 to 30 percent inherent moisture by weight. The heat content of subbituminous coal ranges from 17 to 24 million Btu per ton on a moist, mineral-matter-free basis. The heat content of subbituminous coal consumed in the United States averages 17 to 18 million Btu per ton, on the as-received basis (i.e., containing both inherent moisture and mineral matter).

Substation
A facility used for switching and/or changing or regulating the voltage of electricity. Service equipment, line

transformer installations, or minor distribution or transmission equipment are not classified as substations.

Substrate

The physical material upon which a photovoltaic cell is made.

Subsystem

Any one of several components in a photovoltaic system (i.e., array, controller, batteries, inverter, load).

Sub-Transmission System

A high voltage system that takes power from the highest voltage transmission system, reduces it to a lower voltage for more convenient transmission to nearby load centers, delivering power to distribution substations or the largest industrial plants. Typically operating at voltages from approximately 35 kV to 100 kV.

Successive Approximation Register (SAR)

Used to perform the analog-to-digital conversion in successive steps in many analog-to-digital (ADC) converters.

Suction Head

Additional energy in a re-action turbine hydro system, created by the draught tube channeling outlet water. Inlet pressure, from the water 'pushing' on the turbine runner as it enters, creates ~80% of the energy in a reaction system. Suction head, from the vacuum created by the closed outlet system, 'pulls' on the runner as the water exits the system, adding up to ~20% additional power output to the system.

Sulfation

A condition that afflicts unused and discharged batteries; large crystals of lead sulfate grow on the plate, instead of the usual tiny crystals, making the battery extremely difficult to recharge.

Sulfation (Battery)

The formation of lead sulfate of such physical properties that it is extreemly difficult, if not impossible, to reconvert it to active material.

Sulfur

One of the elements present in varying quantities in coal which contributes to environmental degradation when coal is burned. In terms of sulfur content by weight, coal is generally classified as low (less than or equal to 1 percent), medium (greater than 1 percent and less than or equal to 3 percent), and high (greater than 3 percent). Sulfur content is measured as a percent by weight of coal on an "as received" or a "dry" (moisture-free, usually part of a laboratory analysis) basis.

Summer Peak

The greatest load on an electric system during any prescribed demand interval in the summer.

Summing Amplifier

An op amp that combines several inputs and produces an output that is the weighted sum of the inputs.

Sun Tempering

A sun-tempered building is elongated in the east-west direction, with the majority of the windows on the south side. The area of the windows is generally limited to about 7% of the total floor area. A sun-tempered design has no added thermal mass beyond what is already in the framing, wall board, and so on. Insulation levels are generally high.

Sunspace

A room that faces south, or a small structure attached to the south side of a house.

Super Draw Lead

Also known as a split conductor. Historically bushings offered a draw lead rating of 400 amps, but by using Trench's split conductor in COTA bushings the draw leads now have ratings of 3,000 amps.

Super Ultra Low Emission Vehicle

SULEVs are 90% cleaner than the average new model year car.

Superconducting Magnetic Energy Storage (SMES)

SMES technology uses the superconducting characteristics of low-temperature materials to produce intense magnetic fields to store energy. SMES has been proposed as a storage option to support large-scale use of photovoltaics and wind as a means to smooth out fluctuations in power generation.

Superconductivity

The pairing of electrons in certain materials when cooled below a critical temperature, causing the material to lose all resistance to electricity flow. Superconductors can carry electric current without any energy losses.

Superheterodyne Receiver

A radio receiver that combines a locally generated frequency with the carrier frequency to produce a lower-frequency signal (IF, or intermediate frequency) that is easier to demodulate than the original modulated carrier.

Superposition

The superposition principle states that the results in a circuit due to independent sources can be superposed to give the resultant quantity.

Superstrate

The covering on the sunny side of a photovoltaic (PV) module, providing protection for the PV materials from impact and environmental degradation while allowing maximum transmission of the appropriate wavelengths of the solar spectrum.

Supplementary Insulation

In dependent insulation applied in addition to basic insulation in order to provide protection against electric shock in the event of a failure of basic insulation.

Supplier

A person or corporation, generator, broker, marketer, aggregator or any other entity, that sells electricity to customers, using the transmission or distribution facilities of an electric distribution company.

Supply Mains

Permanently installed power source which may also be used to supply electrical apparatus.

Supply-Side

Technologies that pertain to the generation of electricity.

Supply-Side Management

Steps utilities take to manage their generating and transmission facilities for maximum efficiency.

Surface Acoustic Wave (SAW)

A sound wave that propagates along the surface of a solid and is contained within the solid. SAW devices typically combine compressional and shear components. In Wireless applications, SAW refers to a Surface Acoustic Wave band-pass filter, which exhibits much better out-of-band rejection, but has higher passband ripple and insertion loss.

Surface Flashover

Surface flashover is a breakdown of the medium in which the solid is immersed. The role of the solid dielectric is only to distort the field so that the electric strength of the medium is exceeded.

Surface-Mounted

An inlet designed to be surface mounted on a panel or piece of equipment.

Surge

A transient (or momentary) wave of current, potential, or power in an electric circuit. The word "surge" has different meanings to different engineering communities. To the protection engineer a "surge" is a transient overvoltage with a duration of a few microseconds, i.e., a spike. To others a "surge" is a momentary overvoltage lasting up to a few seconds, a swell.

Surge Arrester

A protective device for limiting surge voltages on equipment by discharging or bypassing surge current. It prevents continued flow of follow through current to earth, and is capable of repeating these functions as specified.

Surge Capacity

The maximum power, usually 3-5 times the rated power, that can be provided over a short time.

Surge Damage

The failure of a device due to a sudden increase in the voltage or current.

Surge Suppressor

A protective device for limiting surge voltages on equipment by discharging or bypassing the surge current to ground.

Surge Withstand

A measure of an electrical device's ability to withstand high-voltage or high-frequency transients of short duration without damage.

Surge Withstand Capability (SWC) Test

The SWC test wave is an

oscillatory wave, frequency range of 1-1.5 MHz, voltage range of 2.5-3 kV crest value of first peak, envelope decaying to 50% of the crest value of the first peak in not less than 6 micro seconds from the start of the wave. The source impedance is from 150-200. The wave is to be applied to a test specimen at a repetition rate of not less than 50 tests per second for a period of not less than two seconds.

Surplus

Excess firm energy available from a utility or region for which there is no market at the established rates.

Susceptance

That part of the admittance that does not consume active power. The imaginary part of admittance. For a pure reactance, it is also the inverse of the reactance.

Swallow Counter

The Swallow Counter is one of the three building blocks (swallow counter, main counter, and dual-modulus prescaler) that constitute the programmable divider commonly used in modern frequency synthesizers.

The swallow counter is used to control the dual-modulus prescaler which is set to either N or (N+1). At the initial reset state, the prescaler is set to a divide ratio of (N+1), but the swallow counter will change this divide ratio to N when it finishes counting S number of cycles.

The Swallow Counter gets its name from the idea that it "swallows" 1 from (N+1) of the dual-modulus prescaler.

Sweeling (Battery)

The swelling or bulging of a battery case that results from cell vents not allowing enough internal pressure to be relieved.

Sweep

Time-dependent information created by the electron beam moving across a CRT screen.

Swell

A momentary overvoltage lasting up to a few seconds.

Switch

A mechanical device capable of making, carrying and breaking current under normal circuit conditions, which may include specified operating overload conditions, and also of carrying for a specified time currents under specified abnormal circuit conditions such as those of short circuit. It may also be capable of making, but not breaking, short circuit currents.

Switch Limit

A switch that is operated by some part or motion of a power-driven machine or equipment to alter the electric circuit associated with the machine or equipment.

Switch Mode

Uses a switching transistor and inductor to control/regulate the charging voltage/current.

Switchboard

A large single panel, frame, or assembly of panels on which are mounted on the face, back, or both, switches, overcurrent and other protective devices, buses, and usually instruments. Switchboards are generally accessible from the rear as well as from the front and are not intended to be installed in cabinets. An assembly of switchgear

with or without instruments, but the term does not apply to groups of local switches in final circuits.

Switched Capacitor Circuit

A circuit methodology, typically implemented in CMOS integrated circuits, that uses clocked switches and capacitors to transfer charge from node to node such that a resistor function is realized. The effective resistance is governed by capacitor size and switching clock frequency.

Switches

Circuit interruption devices used to control the flow of electricity to lights, appliances, and outlets.

Switchgear

A general term covering switching and interrupting devices and their combination with associated control, metering, protective and regulating devices. Also, the assemblies of these devices with associated interrconection, accessories, enclosures and supporting structure, used primarily in connection with the generation, transmission, distribution and conversion of electric power.

Switching Current (in Amps)

The maximum current a reed contact can switch.

Switching Frequency (in Hz)

The maximum frequency at which a reed contact can operate.

Switching Regulator

A voltage regulator that uses a switching element to transform the supply into an alternating current, which is then converted to a different voltage using capacitors, inductors, and other elements, then converted back to DC. The circuit includes regulation and filtering components to insure a steady output. Advantages include the ability to generate voltages beyond the input supply range and efficiency; disadvantages include complexity.

Switching Station

Facility equipment used to tie together two or more electric circuits through switches. The switches are selectively arranged to permit a circuit to be disconnected, or to change the electric connection between the circuits.

Switching Surges

High voltage spikes that occur when current flowing in a highly inductive circuit, or a long transmission line, is suddenly interrupted. As the magnetic field about the inductive conductor collapses, a brief by very high voltage can be generated at the terminal point of the circuit.

Switching Voltage (in Volts)

The maximum voltage a reed contact can switch.

Symmetrical Components

The analysis of an unbalanced three phase system into three balanced components, namely the positive sequence, negative sequence and zero sequence.

Synchronous Digital Hierarchy (SDH)

The ITU-TSS International standard for transmitting information over optical fiber.

Synchronous Machine

A machine which runs at a fixed speed, dependant on the frequency, called the synchronous speed. The machine speed is thus independent of the load.

Synchronous Speed

The speed of the rotating magnetic field created by the primary winding of a rotating electric machine. When the speed of the rotating element matches the speed of the rotating magnetic field, it is said to be rotating at synchronous speed.

Synchronous speed = Frequency x 120 / Number of poles

Synthetic Oil

Oil produced by artificial means and having similar properties to mineral oil.

Tachometer

An instrument for measuring speed (rpm) of a rotating shaft.

Tag Line

A rope used to control the position of equipment being lifted. This is not to be confused with the rope used to actually lift the equipment.

Tail

The proper term is actually Vane, but Tail is commonly used.

Tail Boom

A strut that holds the tail (Vane) to the wind generator frame.

Tail Race

The channel or pipe carrying water from a hydro turbine outlet back to the stream/river it was diverted from.

Tandem

A wallplate with individual gangs arranged vertically one above the other.

Tap

A connection brought out of the winding at some point between its extremities, usually to permit changing the voltage or current ratio.

Taper

Refers to the configuration of the end-to-end resistor. It is usually reported as linear (evenly distributed) or logrithmic.

Tare Loss

Loss caused by a charge controller. One minus tare loss, expressed as a percentage, is equal to the controller efficiency.

T-Body

A device used to terminate main feeder cables operating at medium voltages (4-35KV nominal). T-Bodies are molded from synthetic rubber and are electrically shielded. They are frequently stacked for multiple terminations and are rated at 600 Amps and are classed "Non-Loadbreak".

Tebi (Ti)

Binary multiple prefix corresponding to terabinary or 2^{40} or $(2^{10})^4$ or 1024^4. [IEC 1998]

Technical Ceramic

A ceramic that exhibits a high degree of industrial efficiency through carefully designed microstructures and superb dimensional precision.

Telemetering

The transmission of measuring, alarm and control signals to and from remote station controls and a central monitoring location.

Temper

The softness of a metal; terms such as soft-drawn, dead soft, annealed, and semi-annealed are used to describe tempers used for conductor metals.

Temperature Coefficient (T.C.)

In a capacitor, the rating which determines the change in capacitance corresponding to a given change in operating temperature. It is usually expressed as the change in capacitance per unit of capacitance per degree Celsius.

Temperature Comparator

An integrated circuit with a digital output that indicates whether a measured temperature is above or below a predetermined threshold.

Temperature Compensation

A circuit that adjusts the charge controller activation points depending on battery temperature. This feature is recommended if the battery temperature is expected to vary more than ±5°C from ambient temperature.

Temperature Sensor

Temperature sensor that uses an external diode-connected transistor as the sensing element to measure temperatures external to the sensor (for example, on a circuit board or on the die of a CPU). Generally produces a digital output.

Temperature Switch

A circuit that opens and closes a conductive path based on temperature.

Temporary Magnet

An artificial magnet that loses its magnetism after the magnetizing force is removed. Soft iron is an example of a temporary magnet.

Tensile Strength

The greatest longitudinal force that a substance can bear without tearing apart or rupturing; also called ultimate tensile strength.

Tension

The force in pounds of kilo grams on a conductor installed overhead. Too much tension on an overhead line can contribute to mechanical failure.

Tera (T)

Decimal multiple prefix corresponding to Trillion (US) or 1012.

Terminal Block

An insulating base equipped with terminals for connecting wires.

Terminals

The parts of a battery to which the external electric circuit is connected.

Tesla

Standard unit of magnetic flux density equal to one weber per square meter. The previously used unit of flux density was the gauss, which was equal to one magnetic line per square centimeter.

Testing

The process of verifying the properties of an equipment by a process of application and measurement.

Thermal Breakdown

Heat is generated continuously in electrically stressed insulation by dielectric losses, which is transferred to the surrounding medium by conduction through the solid dielectric and by radiation from its outer surfaces. If the heat generated exceeds the heat lost to the surroundings, the temperature of the insulation increases leading ultimately to thermal breakdown if a stable temperature is not reached.

Thermal Control Circuit

Circuit to monitor and control the temperature of

something. For example the integrated temperature controller in Intel's processors.

Thermal Electric
Electric energy derived from heat energy, usually by heating a working fluid, which drives a turbogenerator.

Thermal EMF
The EMF generated by a reed contact when the reed contact unit is subjected to a temperature differential. For most reed switches, this is in the region of 40 micro volts per degree C.

Thermal Expansion
The expansion of a material when subjected to heat.

Thermal Management
The use of various temperature monitoring devices and cooling methods, such as forced air flow, within a processor or FPGA-based system, to control overall temperature of ICs and internal cabinet temperatures.

Thermal Mass
Materials that store heat within a sunspace or solar collector.

Thermal Mechanism
The thermal mechanism

uses a heat sensitive bimetal element to operate the circuit breaker in the event of an overcurrent.

Thermal Monitor
The integrated thermal control system used in Intel's processor devices.

Thermal Ohm
The thermal ohm is the resistance of a path through which a temperature difference of 1 0C produces a heat flow of 1 watt.

Thermal Overload Protector
Device which protects motor windings from excessive temperature by opening a set of contacts.

Thermal Resistance
Opposition to the flow of heat.

Thermal Resistivity
The thermal resistivity is the temperature drop in degree celsius produced by the flow of 1 watt between the opposite faces of a metre cube of the material.

Thermal Runaway
A condition whereby a cell on charge or discharge will destroy itself through inter-

nal heat generation caused by high overcharge or high rate of discharge or other abusive conditions.

Thermal Shock

Thermal shock is the effect of heat or cold applied at such a rate that non-uniform thermal expansion or contraction occur within a given material or combination of materials. In connectors, the effect can cause inserts and other insulation materials to pull away from metal parts.

Thermal Shutdown

Deactivating a circuit when a measured temperature is beyond a predetermined value.

Thermal Storage Walls (Masonry Or Water)

A thermal storage wall is a south-facing wall that is glazed on the outside. Solar heat strikes the glazing and is absorbed into the wall, which conducts the heat into the room over time. The walls are at least 8 in thick. Generally, the thicker the wall, the less the indoor temperature fluctuates.

Thermionic Emission

The liberation of electrons from a solid metal as a result of heat (thermal energy).

Thermistor

A resistive device used for temperature sensing that is composed of metal oxides formed into a bead and encapsulated in epoxy or glass. A typical thermistor has a positive temperature coefficient; that is, resistance increases dramatically and non-linearly with temperature. Though less common, there are negative temperature coefficient thermistors.

Thermochron

A device which measures and records (logs) temperature. Trademark of Dallas Semiconductor.

Thermocouple

A temperature sensor formed by the junction of two dissimilar metals. A thermocouple produces a voltage proportional to the difference in temperature between the hot junction and the lead wire (cold) junction.

Thermophotovoltaic (TPV) device

A device that converts secondary thermal radiation,

re-emitted by an absorber or heat source, into electricity; The device is designed for maximum efficiency at the wavelength of the secondary radiation.

Thermoplastic
A plastic compound that will soften and melt with sufficient heat. Thermoplastic insulation compounds are used to manufacture certain types of electrical cables.

Thermoset
A plastic compound that will not remelt. Thermoset insulation compounds are used to manufacture certain types of cables.

Thermosetting Plastics
Plastics which, having once been subjected to heat and pressure lose their plasticity.

Thermostat
Circuit that indicates whether a measured temperature is above or below a particular temperature threshold or trip point. Used for thermal protection and simple temperature control systems.

Thevenin's Theorem
States that a linear two-ter-minal circuit can be replaced by an equivalent circuit consisting of an equivalent voltage source and a series equivalent impedance.

Thick-Crystalline Materials
Semiconductor material, typically measuring from 200-400 microns thick, that is cut from ingots or ribbons.

Three Phase
Three-phase refers to an electric power system having at least three conductors carrying voltage waveforms that are 2ð/3 radians (120°,1/3 of a cycle) offset in time. Electric utilities generate three phase power and transmit it to load centers where it may be consumed at Three Phase or Single Phase.

Three-way Switch
A pair of switches wired to control the same fixture or group of fixtures.

Through Fault Current
The current flowing through a protected zone to a fault beyond that zone.

Through-Hole
A method for mounting components on a printed

circuit board (PCB) in which pins on the component are inserted into holes in the board and soldered in place.

Throughput

A general term used when defining the rate of data transfer over a particular medium, such as a wireless network or a phone line.

Thrust

In a wind generator, wind forces pushing back against the rotor. Wind generator bearings must be designed to handle thrust or else they will fail.

Thrust Bearing

A bearing that is designed to handle axial forces along the centerline of the shaft. In a wind generator, this is the force of the wind pushing back against the blades.

Thumper

A high voltage device used to locate an underground cable fault. The device applies a high voltage to the faulted cable with a resulting discharge to ground at the location of the fault. When the discharge occurs, there is an audible "Thump" which is used to locate the fault. Hipotronics manufactures "Thumpers".

Thunder

The sound that follows a flash of lightning and is caused by the sudden expansion of the air in the path of the electrical discharge. This explosive heat produces a massive, deafening shock wave-thunder.

Thyristor

A solid-state switching device for semiconductors to convert AC current in one of two directions controlled by an electrode. A four-layer semiconductor device that acts as a latch.

Tie

A wire device that connects a conductor to an insulator. Factory formed ties are manufactured by Preformed Line Products Company.

Tilt Angle

The angle at which a photovoltaic array is set to face the sun relative to a horizontal position. The tilt angle can be set or adjusted to maximize seasonal or annual energy collection.

Tilt-Up

A tower that is hinged at the

base and tilted up into position using a gin pole and winch or vehicle. Wind turbines on tilt-up towers can be serviced on the ground, with no climbing required.

Time Division Duplex (TDD)

Time Division Duplex, the second variation of W-CDMA especially suited to indoor environments where there is a need for high traffic density.

Time Division Multiple Access (TDMA)

A method of digital wireless-communications transmission. TDMA allows many users to access (in sequence) a single radio-frequency channel without interference, because it allocates unique time slots to each user within each channel.

Time Division Multiplexing (TDM)

A scheme in which numerous signals are combined for transmission on a single communications line or channel. Each signal is broken into many segments, each having very short duration.

Timer

A switch with an integral mechanism or electronic circuit that can be set to switch an electrical load ON at a predetermined time.

Tin Oxide

A wide band-gap semiconductor similar to indium oxide; used in heterojunction solar cells or to make a transparent conductive film, called NESA glass when deposited on glass.

Tiny Internet Interface (TINI)

Tiny Internet Interface: The industry's smallest web server, TINI is a microcontroller that includes the facilities necessary to connect to the Internet. The platform is a combination of broad based I/O, a full TCP/IP stack and an extensible Java runtime environment that simplifies development of network-connected equipment.

Tip

The end of a wind generator blade farthest from the hub.

Tip Speed Ratio

The ratio of how much faster than the windspeed that the blade tips are moving. Abbreviation: TSR

Tipping Fee

A credit received by municipal solid waste companies for accepting and disposing of solid waste.

TN System

A system having one or more points of the source of energy directly earthed, the exposed conductive parts of the installation being connected to that point by protective conductors.

TNS System Of Earthing

In this earthing system, the supplier provides separate Neutral and Protective conductors throughout the system. The Protective Conductor is connected to the neutral of the source. All exposed conductive parts of a consumer's installation are connected to the Protective Conductor provided by the supplier via. the main earthing terminal of the consumer's installation.

Toggle

A switch with a lever-type actuator that makes or breaks switch contact as its position is changed.

Tonne

Metric tonne. An unit of weight. 1 tonne = 1000 kg

Toroid

This refers to a circular, donut shaped core used in transformers Toriod cores are generally molded from powdered iron or wound with silicon steel strips.

Torr

Unit of pressure. 1 torr = 1 mm of Hg = 133.3 pa

Total AC Load Demand

The sum of the alternating current loads. This value is important when selecting an inverter.

Total Harmonic Distortion

The measure of closeness in shape between a waveform and it's fundamental component.

Totem Pole

A standard CMOS output structure where a P-channel MOSFET is connected in series with an N-Channel MOSFET and the connection point between the two is the output. The P-FET sits on top of the N-FET like a "totem pole." Both gates are driven by the same signal. When the signal is low, the P-FET is on; when the signal is high, the N-FET is on. This creates a push-pull output using just two transistors.

Touch Voltage

The potential difference between the ground potential rise (GPR) and the surface potential at the point where a person is standing where at the same time having his hands in contact with a grounded structure. GPR defined as the maximum voltage that a station grounding grid may attain relative to a distant grounding point assumed to be at the potential of remote earth. The touch voltage could be from hand to hand also.

Tower

A structure (usually steel) found along transmission lines which is used to support conductors.

Track And Accent Lighting

Condition specific lighting that meets special lighting requirements, providing variable lighting degrees of light and may distribute light in multiple directions.

Tracker

Any device used to direct a PV array toward the sun.

Tracking

Tracking is the formation of a permanent conducting path across a surface of the insulation, and in most cases the conduction (carbon path) results from degradation of the insulation itself leading to a bridge between the electrodes. Tracking occurs in organic materials.

Tracking Array

A photovoltaic (PV) array that follows the path of the sun to maximize the solar radiation incident on the PV surface. The two most common orientations are (1) one axis where the array tracks the sun east to west and (2) two-axis tracking where the array points directly at the sun at all times. Tracking arrays use both the direct and diffuse sunlight. Two-axis tracking arrays capture the maximum possible daily energy.

Tracking PV Array

PV array that follows the path of the sun to maximize the solar radiation incident on the PV surface. The two most common orientations are (1) one axis where the array tracks the sun east to west and (2) two-axis tracking where the array points directly at the sun at all times. Tracking arrays use both the direct and diffuse sunlight. Two-axis tracking arrays capture the maximum possible daily energy.

Tracking system
A solar array mount that rotates the panels to face toward the sun throughout the day.

Trailing Edge
The edge of a blade that faces away from the direction of rotation.

Transceivers
A device that contains both a transmitter and receiver.

Transco
A for profit Power Transmission Company.

Transconductance
The gain of a transconductance amplifier (an amp in which the a change in input voltage causes a linear change in output current). The basic gain of vacuum tubes and FETs is expressed as transconductance. It is represented with the symbol g_m. The term derives from "transfer conductance" and is measured in siemens (S), where 1 siemens = 1 ampere per volt. It was formerly measured as "mho" (ohm spelled backwards).

Transconductance Amplifier
An amplifier that converts a voltage to a current. Also known by several other terms. One synonym is OTA, or operational transconductance amplifier, a term that marries the terms transconductance amplifier and operational amplifier. The term derives from "transfer conductance" and is measured in siemens (S), where 1 siemens = 1 ampere per volt. It is represented with the symbol g_m. The basic gain of vacuum tubes and FETs is expressed as transconductance.

Transducer
A device to condition and transform a specific physical quantity to a specific variable output electrical signal proportional to the input signal. Typical inputs include variable pressure, level, voltage or current. A transducer must be specifically designed to be compatible with the input/output

requirements of the total system.

Transducer Error

The actual value of the output minus the intended value of the output expressed algebraically.

Transducer Factor

The product of the current transformer ratio (CTR) and the voltage transformer ratio (VTR). Also called the power ratio.

Transducer with Live Zero

A transducer which gives a predetermined output other than zero when the measurand is zero.

Transfer

To move electric energy from one utility system to another over transmission lines.

Transfer Switch

An electronic device that under certain conditions will disconnect from one power source and connect to another power source.

Transferred Voltage

This is a special case of the touch voltage where the voltage is transferred into or out of the station by a conductor grounded at a remote point or at the station ground, respectively.

Transformation

Change of one form variable or substance into another.

Transformer Insulation

This is the material that is used to provide electrical insulation between transformer windings at different voltage levels and also between the energized parts and the metal tank of the transformer. Generally, for large transformers used in power applications, this is a combination of Kraft paper and mineral oil; however, in indoor building applications, transformers also may be insulated with plastic materials.

Transformer Ratio

When used in reference to Instrument Transformers, this is simply the ratio of transformation of one or more transformers used in the circuit. If both Cts and VTs are included, the transformer ratio is the product of the CT and the VT. For example, assume a VT with a ratio 10:1 and a CT with a ratio of 20:1, then the combined transformer ratio is 10 x 20 = 200:1.

Transient

A phenomenon of a non-repetitive nature caused by a sudden change in conditions that persist for a relatively short time after the change.

Transient Response

The ability of a power conditioner to respond to a change. Transient step load response is the ability of a power conditioner to maintain a constant output voltage when sudden load (current) changes are made.

Transient Voltage Suppressor (TVS)

Semiconductor device designed to protect a circuit from voltage and current transients. Typically implemented as a large silicon diode operating in avalanche mode to absorb large currents quickly.

Transimpedance Amplifier

An amplifier which converts a current to a voltage. It is a familiar component in fiber-communications modules. The unit for transresistance is the ohm.

Transistor

A basic solid-state control device which allows or disallows current flow between two terminals, based on the voltage or current delivered to a third terminal.

Usually built from silicon but can be constructed from other semiconductor materials. There are two major types: The FET (field-effect transistor) and the bipolar junction transistor (BJT).

The first transistor was invented in 1947 at Bell Labs by Michael John Bardeen, Walter Brattain and William Shockley.

Transmission

The movement or transfer of electric energy over an interconnected group of lines and associated equipment between points of supply and points at which it is transformed for delivery to consumers, or is delivered to other electric systems. Transmission is considered to end when the energy is transformed for distribution to the consumer.

Transmission and Distribution (T&D) Losses

Losses the result from the friction that energy must

overcome as it moves through wires to travel from the generation facility to the customer. Because of losses, the demand produced by the utility is greater than the demand that shows up on the customer bills.

Transmission Charge

Part of the basic service charges on every customer's bill for transporting electricity from the source of supply to the electric distribution company. Public Utility Commissions regulate retail transmission prices and services. The charge will vary with source of supply.

Transmission Lines

Heavy wires that carry large amounts of electricity over long distances from a generating station to places where electricity is needed. Transmission lines are held high above the ground on transmission towers.

Transmission System

Normally, the highest voltage network of an electric utility system. This is the portion of the system that carries high power over the longest distances. Typically operating at voltages in ex-

cess of 100 kV, and most usually at 200 kV and above.

Transverse Mode Noise

(Normal mode)- An undesirable voltage which appears from line to line of a power line.

Trapezoidal Wire (TW)

1) A thermoplastic insulated, moisture resistant conductor designed for use in wet or dry locations and an operatating temperature of up to 60 degrees Celsius. 2) Trapezoidal Wire. Built as ACSR-TW or ACSS-TW, Trapezoidal Wire uses trapezoidal formed strands in its construction to reduce overall diameter of the finished cable.

Trash Rack

A mesh or set of bars at the inlet of a hydro system that catches leaves, branches and other debris, and prevents it from obstructing or damaging the turbine.

T-Rated

A switch specially designated with the letter "T" in its rating that is rated for controlling tungsten filament lamps on direct current (DC) or alternating current (AC) circuits.

Travelling Waves

Voltage waveforms which effectively travel along a line without a significant change in waveshape.

Traverse-Mode Noise

Often used as a synonym for normal-mode noise, it more clearly relates to noise that is the result of the conversion of common-mode noise to normal-mode noise after it passes through a transformer.

Tree of a Network

A graph of the network with some of the links removed in such a way so as to leave all the nodes connected together by the graph, but so as not to have any loop left in the network.

Tree Retardant Cross Linked Polyethylene (TRXLP)

A thermoset plastic compound that is used for insulation of wire and cable contaning an anti-treeing compound.

Tree Wire

A type of Overhead Distribution Wire that is insulated for momentary contact with tree branches and used as a primary voltage conductor.

Treeing

Water treeing is a form of cable insulation degradation where micochannels, that often appear as a tree-like structure in the insulation, develop due to a complex interaction of water, electrical stress, impurities and imperfections. The tree-like channels grow slowly over time, weakening the insulation and, in some cases, lead to cable failure.

Trending

Analysis of the change in measured data over at least three data measurement intervals.

Triac

A thyristor that can conductor in both directions. Because of this, it is useful for controlling alternating current. It is the equivalent of two SCRs in parallel with opposite polarities.

Triangle of Forces

If three forces acting at the same point can be represented in magnitude and direction by the three sides of a triangle, taken in order, they will be in equilibrium.

Trickle Charge

With the trickle charging process, the battery receives a constant voltage feeding a low current. Constant use of this method dries the electrolyte and corrodes the plate, reducing potential battery service life by up to 50 percent.

Trigger

The signal used to initiate a sweep on an oscilloscope and determine the beginning point of the trace.

Trigger Holdoff

A front-panel control that inhibits the trigger circuit from looking for a trigger for some specified time after the end of the trace.

Trigger Level

The instantaneous level that a trigger source signal must reach before a sweep is initiated by the trigger circuit.

Triplex

A receptacle with a common mounting means which accepts three plugs.

True Value

Average value of the infinite number of measurements, when the average deviation tends to become zero..

TT System of Earthing

An earthing system where all exposed conductive parts of an installation are connected to an earth electrode provided by the consumer which is electrically independent of the Source earth.

Tubular Plate (Battery)

A positive plate which is composed of asembly of porous tubes of perforated metal or tissure with or without a central current collector spine. The active material is placed within the tube.

Tungsten Halogen Lamp

A gas-filled tungsten halogen lamp containing a certain proportion of halogens.

Tunnel Effect

The passage of an electron through a narrow potential barrier in a semiconductor, despite the fact that, according to classical theory, the electron does not possess sufficient energy to surmount the barrier.

Tunneling

Quantum mechanical concept whereby an electron is found on the opposite side of an insulating barrier without having passed through or around the barrier.

Turbine

A machine for generating rotary mechanical power from the energy of a stream of fluid (such as air, water, steam, or hot gas). Turbines convert the kinetic energy of fluids to mechanical energy. It generally consists of a series of curved vanes emanating from an axis that is turned by forcing the fluid past the vanes.

Turbine Generator

The combination of a turbine and a generator working together to produce power.

Turgo Turbine

An impulse turbine that can provide higher power output in some high head hydro sites than a Pelton style. The Turgo runner design allows for more efficient escape of 'used' water, and a larger water jet, for improved power production in sites with high enough flow rates. Turgo runners are generally more expensive than Pelton style, due to more difficult manufacturing processes.

Turn Ratio

The ratio of the number of turns in the high voltage winding to that in the low voltage winding.

Twist

In a wind generator blade, the difference in Pitch between the blade root and the blade tip. Generally, the twist allows more Pitch at the blade root for easier Startup, and less Pitch at the tip for better high-speed performance.

Twisted Pair

Telephone companies commonly run twisted pairs of cooper wires to each customer household. The pairs consist of two insulated cooper wires twisted into a spiral pattern. These wires are capable of transferring both voice as well as data..

Two-Axis Tracking

A system capable of rotating independently about two axes (e.g., vertical and horizontal) and following the sun for maximum efficiency of the solar array.

Two-Port Network

An electrical network with two separate ports for input and output.

Two-Way Switch

A two-way switch is one which enables operation of a lamp from two positions, such as at the top and bottom of a staircase and at the ends of a long corridor.

Type Tests

These tests are done to ensure that the particular design is suitable for a specific purpose. They are normally done either at design stage, or when a purchaser (of large orders) requires them to be done.

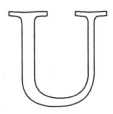

Ultra High Voltage (UFV)
Electric systems in which the Root-Mean-Square ac voltage exceeds 800,000 volts.

Ultra Low Emission Vehicle
ULEVs are 50% cleaner than the average new model year car.

Ultrahigh Voltage Transmission
Transporting electricity over bulk-power lines at voltage greater than 800 kilovolts.

Ultraviolet
Electromagnetic radiation in the wavelength range of 4 to 400 nanometers. vacuum evaporation-The deposition of thin films of semiconductor material by the evaporation of elemental sources in a vacuum.

Ultra-Wideband (UWB)
Ultra-Wideband (UWB) is a communications technology that employs a wide bandwidth (typically defined as greater than 20% of the center frequency or 500MHz). UWB is usually used in short-range wireless applications but can be sent over wires. Ultra-Wideband advantages are that it can carry high data rates with low power and little interference.

UWB is the modern version of older "impulse" technologies which are generated by very short pulses (impulse waveforms). They were called "carrier-free" or "baseband" because the energy is so widespread in the frequency domain that there is no discernible carrier frequency.

For a crude example, connect a metal file to one terminal of a battery and a wire to the other terminal. Brush the wire across the teeth of the file and note that the electrical noise can be heard on a radio tuned to just about any frequency.

The FCC authorizes UWB between 3.1 and 10.6GHz (but is not likely to approve devices that rely on a file and a wire.)

Unbalanced Loads

Refers to an unequal loading of the phases in a polyphase system.

Unbalanced Three Phase System

A three phase system in which either the supply or the load or both are not fully balanced.

Unbundled Services

Unbundling refers to the process of disaggregating electric utility services into basic components and offering each component for sale, with separate charges for each component. Some services may be regulated and others are subject to competition. Some services may be mandatory (meaning all customers must buy them) and others may be optional (customers have a choice whether or not to buy them, and often from whom).

Once unbundled, various electric utility and competitive services may be listed on customer bills as separate line items.

Unbundling

Disaggregating electric utility service into its basic components and offering each component separately for sale with separate rates for each component. For example, generation, transmission and distribution could be unbundled and offered as discrete services.

Uncertainties

Uncertainties are factors over which the utility has little or no foreknowledge, and include load growth, fuel prices, or regulatory changes. Uncertainties are modeled in a probabilistic manner. However, in the Detailed Workbook, you may find it is more convenient to treat uncertainties as "unknown but bounded" variables without assuming a probabilistic structure. A specified uncertainty is a specific value taken on by an uncertainty factor (e.g. 3 percent per year for load growth). A future uncertainty is a combination of specified uncertainties (e.g. 3 percent per year load growth, 1 percent per year real coal and oil price escalation, and 2.5 percent increase in housing starts).

Underground Feeder (UF)

May be used for photovoltaic array wiring if sunlight resistant coating is specified; can be used for interconnecting balance-of-system components but not recommended for use within battery enclosures.

Underground Residential Distribution

Refers to the system of electric utility equipment that is installed below grade.

Underground Service Entrance (USE)

May be used within battery enclosures and for interconnecting balance-of-systems.

Underground Utility Structure

An enclosure for use underground that may be either a handhole or manhole.

Under-Voltage

Like a sag, but for a longer period of time: over 2.5 seconds.

Underwriters' Laboratories (UL)

Underwriters' Laboratories is a nonprofit organization that tests electrical devices to assure their compliance with the NEC.

Unidirectional Unit

Allows inputs to be measured in one direction only. The stated output range indicates the minimum and maximum input levels.

Uniform System of Accounts

Prescribed financial rules and regulations established by the Federal Energy Regulatory Commission for utilities subject to its jurisdiction under the authority granted by the Federal Power Act.

Unijunction transistor UJT

This low power transistor is useful in electronic timing, waveshaping and control applications.

Uninterruptible Power Supply

A device that provides a constant regulated voltage output in spite of interruptions of the normal power supply. It includes filtering circuits and is usually used to feed computers or related equipment which would otherwise shutdown on brief power interruptions. Abbreviated UPS.

UniqueWare
A unique identification technique

UniqueWare Serialized
A factory-programming service for 1-Wire EPROM chips with customer-specified data. Service provides one serialization file for customers to create identifiers in silicon.

Unit Cell
The basic structural unit of a crystal structure.

Unit Electrical Relay
A single relay that can be used alone or in combinations with others.

Unit Energy Consumption (UEC)
The annual amount of energy that is used by the electrical device or appliance.

Unit Impulse
A function with unit integral area, which is zero everywhere except at zero time where it is infinite.

Unit interval (UI)
Unit interval (used to de-

scribe jitter generation); user information; user interface.

Unit of Current
Practical unit of current is the ampere, which is the current produced by a pressure of one volt in a circuit having a resistance of one ohm.

Unit of Pressure
The volt, or pressure which will produce a current of one ampere against a resistance of one ohm.

Unit of Resistance
The ohm, which is the resistance that permits a flow of one ampere when the impressed pressure is one volt.

Unit Protection
A protection system that is designed to operate only for abnormal conditions within a clearly defined zone of the power system.

Unit Ramp
A function which is zero for negative time, and unit slope for positive time.

Unit Step
A function with zero magnitude for negative time and unit magnitude for positive time.

Universal Asynchronous Receiver-Transmitter (UART)
An IC that converts parallel data to serial, for transmission; and converts received serial data to parallel data.

Universal Bushing Well
This 200 amp rated component is used as part of a system to terminate medium voltage cables to transformers, switchgear and other electrical equipment. Universal Bushing Wells are manufactured by the Elastimold Division of Thomas & Betts.

Universal Mobile Telecommunications System (UMTS)
A future mobile communications system which, among other features, will offer direct connection between terminals and satellites.

Universal Resource Locator (URL)
This is the method of addressing on the web. They include the file transfer protocol (ftp) and the hyper text transfer protocol (http).

Universal Serial Bus (USB)
A standard port that enables you to connect external devices (such as digital cameras, scanners, keyboards, and mice) to computers. The USB standard supports data transfer at three rates: low speed (1.5MBps), full speed (12Mbps) and high speed (480 MBps). Mbps=million bits per second.

Universal Service
Electric service sufficient for basic needs (an evolving bundle of basic services) available to virtually all members of the population regardless of income.

Universal Taps
A combination of six primary voltage taps consisting of four 2-1/2% FCBN and two 2-1/2 FCAN, covering 15% voltage range.

Unrestricted Protection
A protection system which has no clearly defined zone of operation and which achieves selective operation only by time grading.

Unserved Energy
The average energy that will be demanded but not served during a specified period due to inadequate available generating capacity.

Upconverters

Device which provides frequency conversion to higher frequency, e.g., digital broadcast-satellite applications.

Upgrade

Replacement or addition of electrical equipment resulting in increased generation or transmission capability.

Uprate

An increase in the rating or stated measure of generation or transfer capability.

Useful Thermal Output

The thermal energy made available for use in any industrial or commercial process, or used in any heating or cooling application, i.e., total thermal energy made available for processes and applications other than electrical generation.

User

Authority responsible for the use and maintenance of equipment.

Utilisation

The process of exploiting (using) electrical energy for various purposes.

Utility

A regulated entity which exhibits the characteristics of a natural monopoly. For the purposes of electric industry restructuring "utility" refers to the regulated, vertically-integrated electric company. "Transmission utility" refers to the regulated owner/operator of the transmission system only. "Distribution utility" refers to the regulated owner/operator of the distribution system which serves retail consumers.

Utility Distribution Companies

The entities that will continue to provide regulated services for the distribution of electricity to customers and serve customers who do not choose direct access. Regardless of where a consumer chooses to purchase power, the customer's current utility, also known as the utility distribution company, will deliver the power to the consumer's home, business, or farm.

Utility-Earned Incentives

Costs in the form of incentives paid to the utility for achievement in consumer participation in DSM programs. These financial incentives are intended to influence the utility's consideration of DSM as a resource option by addressing cost recovery, lost revenue, and profitability.

Utility-Interactive Inverter

An inverter that can function only when tied to the utility grid, and uses the prevailing line-voltage frequency on the utility line as a control parameter to ensure that the photovoltaic system's output is fully synchronized with the utility power.

Utilization Factor

The ratio of the maximum demand of a system or part of a system to the rated capacity of the system or part of the system.

Vacancy

A normally occupied lattice site from which an atom or ion is missing.

Vacuum Circuit Breakers

Circuit breakers, normally applied at medium voltages, that use vacuum interrupters to extinguish the electrical arc & shut-off flowing current.

Vacuum Evaporation

Method of depositing thin coatings of substance by heating it in a vacuum system.

Vacuum Interrupter

A sealed "bottle" containing contacts of a switch inside a very high vacuum. When the contacts are parted in the vacuum, as there is no gas in the bottle to ionize, the current flow is quickly extinguished.

Vacuum Tube

A form of electron tube in which the envelope contains a vacuum, as opposed to a gas-filled electron tube, in which gases are pumped into the envelope after the air is removed.

Vacuum Zero

The energy of an electron at rest in empty space; used as a reference level in energy band diagrams.

Valence Band

The highest energy band in a semiconductor that can be filled with electrons.

Valence Electrons

The outer shell of an atom is known as the valence shell. Any electrons located in the outer shell of an atom are known as valence electrons. The valence shell of an atom cannot hold more than eight electrons. It is the valence electrons that are primary concern in the study of electricity, because it is these that explain much of electrical theory. A conductor for instance, is generally made from a material that contains one or two valence electrons. Atoms with one or two valence electrons are unstable and can be made to give up these electrons with little effort. Conductors are materials that permit electrons to flow through them easily. When an atom has only one or two valence electrons, these electrons are loosely held by the atom and are easily given up for the current flow. Silver, copper, gold, and aluminum all contain one valence electron and are excellent conductors of electricity. Silver is the best natural conductor of electricity, followed by copper, gold, and aluminum.

Valence Level Energy/Valence State

Energy content of an electron in orbit about an atomic nucleus. Also called bound state.

Valence State; Valence Level Energy, Bound State

Energy content of an electron in orbit about an atomic nucleus.

Valley Filling

Valley filling is a form of load management that increases or builds, off-peak loads. This load shape objective is desirable if a utility has surplus capacity in the off-peak hours. If this strategy is combined with time-or-use rates, the average rate for electricity can be lowered.

Valve Regulated Sealed Cell (Battery)

A battery in which the cells are closed but have a valve which allows the excape of gas if the internal pressure exceeds a predetermined value (pressure).

van der Waals Bond

A secondary interatomic bond between adjacent molecular dipoles, which may be permanent or induced.

Vane

A large, flat piece of material used to align a wind turbine rotor correctly into the wind. Usually mounted vertically on the tail boom. Sometimes called a Tail.

Vapor Barrier

Also called a vapor retarder, this is a material that retards the movement of water vapor through a building element (such as walls, floors, and ceilings) and prevents metals from corroding and insulation and structural wood from becoming damp.

Vapor Phase

In the core-type transformer, the core-and-coil assembly is independent of the tank, so that the assembly is allowed to completely dry. When drying the core-and-coil assembly, vapor phase drying method is used, in which special oil vapor is sprayed on the assembly to utilize latent heat produced when the oil vapor condenses. Since heating is effected deep inside evenly and quickly, the assembly can be completely dries without causing damage to the insulation.

Variable Capacitor

A capacitor in which the capacitance can be varied by some mechanical means.

Variable Contact Resistance (VCR)

The difference between lowest and highest static CR readings out of a set consequent test cycles.

Variable Cost

Costs, such as fuel costs, that depend upon the amount of electric energy supplied.

Variable Frequency Drive (VFD)

An electronic device that controls the speed and direction of a motor by changing the voltage and frequency of the power to the motor.

Variable Pitch

A type of wind turbine rotor where the attack angle of the blades can be adjusted either automatically or manually.

Variable Prices

Prices that vary frequently. Prices that are not stable.

Variable Torque Motor

A multi-speed motor in which the rated horsepower varies as the square of the synchronous speeds.

Varistor

A voltage-dependent variable resistor. Normally used to protect sensitive equipment from power spikes or lightning strikes by shunting the energy to ground.

Vector Group Compensation

A feature of digital and numerical relays that compensates for the phase angle shift that occurs in transformers due to use of dissimilar winding connections. For example transformers connected in delta/star.

Veiling Luminance

A luminance superimposed on the retinal image which reduces its contrast. It is this veiling effect produced by bright sources or areas in the visual field that results in reduced visual performance and visibility.

Vent

A normally sealed mechanism that allows for the controlled escape of gases from within a cell.

Vent Cap (Battery)

The plug on top of a cell that can be removed to check and change the level of the electrolyte.

Vent Valve (Battery)

A normally sealed mechanism which allows the controlled excape of gasses from within a cell.

Vented Cell

A battery designed with a vent mechanism to expel gases generated during charging.

Ventilated

Provided with a means to permit circulation of air sufficient to remove an excess of heat, fumes, or vapors.

Venting (Battery)

The release of gas from a cell, either controlled (through a vent) or accidental.

Vertical Integration

An arrangement whereby the same company owns all the different aspects of making, selling, and delivering a product or service. In the electric industry, it refers to the historically common arrangement whereby a utility would own its own generating plants, transmission system, and distribution lines to provide all aspects of electric service.

Vertical Multijunction (VMJ) Cell

A compound cell made of different semiconductor materials in layers, one above the other. Sunlight entering the top passes· through successive cell barriers, each of which converts a separate portion of the spectrum into electricity, thus achieving greater total conversion efficiency of the incident light. Also called a multiple junction cell.

Very High Data-Rate Digital Subscriber Line (VDSL)

A method for delivering high-speed digital services on the standard twisted pair used for voice phone lines. VDSH operates at data rates from 12.9Mbps to 52.8Mbps.

Very High Frequency (VHF)

Radio frequencies in the range 30 MHz to 300 MHz.

Very Large-Scale Integration (VLSI)

Very large-scale integration (VLSI) refers to an IC or technology with many devices on one chip. The question, of course, is how one defines "many."

The term originated in the 1970s along with "SSI" (small-scale integration), "LSI" (large-scale), and several others, defined by the number of transistors or gates per IC. It was all a bit silly since improving technology obviously makes numerical definitions meaningless over time. And it varies by industry — a VLSI analog part is quite different from a VLSI digital logic part or a VLSI memory part. Eventually, the pundits began trying terms like "ULSI" (ultra-large-scale). Engineers, meanwhile, ignored it all and spent their time building better devices instead of making up new words for them.

The terms LSI and VLSI are now usually used as general terms, referring to a product or technology that subjectively has more devices than typical products in the category. Maxim/Dallas Semiconductor has observed a technical trend in analog and mixed signal toward increasing complexity. Many of our parts include complex control, such as the MAXQ microcontroller core, with many times more devices than most analog parts.

Virtual Server

The part of a server that functions as if it were a separate, dedicated server. Each virtual server can run its own operating system and applications and even be networked with other virtual servers on the same machine. For instance, web hosting companies use virtual servers to house multiple clients' web sites on one server.

Viscoelasticity

A type of deformation exhibiting the mechanical characteristics of viscous flow and elastic deformation.

Viscosity

The ratio of the magnitude of an applied shear stress to the velocity gradient that it produces.

Visual Corona

Visible signs (usually a bluish glow) of the presence of corona, which occurs at a higher electric field than is necessary for the inception of corona.

Volatile Memory

Memory requiring electrical power to keep information stored.

Volt

A unit of electrical pressure. It measures the force or push of electricity. Volts represent pressure, correspondent to the pressure of water in a pipe. A volt is the unit of electromotive force or electric pressure analogous to water pressure in pounds per square inch. It is the electromotive force which, if steadily applied to a circuit having a resistance of one ohm, will produce a current one ampere.

Volt Ampere

A unit of apparent power equal to the mathematical product of a circuit voltage and amperes. Here, apparent power is in contrast to real power. On ac systems the voltage and current will not be in phase if reactive power is being transmitted. Usually abbreviated VA.

Voltage

Electrical pressure, the force which causes current to flow through a conductor. Voltage must be expressed as a difference of potential between two points since it is a relational term. Connecting both voltmeter leads to the same point will show no voltage present although the voltage between that point and ground may be hundred or thousands of volts. This is why most nominal voltages are expressed as "phase to phase" or "phase to neutral". The unit of measurement is "volts". The electrical symbol is "e".

Voltage At Maximum Power (Vmp)

The voltage at which maximum power is available from a module. [UL 1703]

Voltage Bands

Band I

Band I covers: Installations where protection against electric shock is provided under certain conditions by the value of voltage;
—installations where the voltage is limited for operational reasons (e.g. telecommunications, signalling, bell, control and alarm installations). Extra low voltage (ELV) will normally fall within voltage band I.

Band II: Band II contains the voltages for supplies to household, and most commercial and industrial installations. Low voltage (LV) will normally fall within voltage band II. Band II voltages do not exceed 1000 V a.c. rms or 1500 V d.c.

Voltage Class

The general strength of electrical insulation on a device, determining the maximum continuous voltage that can be applied between the conducting parts and ground potential, without damaging the insulation.

Voltage Controlled Crystal Oscillator (VCXO)

An oscillator that uses a crystal to establish its frequency but will vary its frequency as an analog control voltage varies.

Voltage Controlled, Temperature Compensated Crystal Oscillator (VCTCXO)

A TCXO which offers the ability to control the oscillation frequency with an analog voltage

Voltage Distortion

Any distortion in the voltage from a perfect sine wave.

Voltage Doubler

A capacitor charge pump circuit which produces an output voltage which is twice the input voltage.

Voltage Drop

The loss of voltage between the input to a device and the

output from a device due to the internal impedance or resistance of the device. In all electrical systems, the conductors should be sized so that the voltage drop never exceeds 3% for power, heating, and lighting loads or combinations of these. Furthermore, the maximum total voltage drop for conductors for feeders and branch circuits combined should never exceed 5%.

Voltage Margining

Setting the output voltage higher or lower than the nominal voltage so that the output voltage remains within the specification during all load conditions.

Voltage Protection

Many inverters have sensing circuits that will disconnect the unit from the battery if input voltage limits are exceeded.

Voltage Rating

The normal voltage to be applied to an electrical device to provide for proper operation.

Voltage Reduction

Any intentional reduction of system voltage by 3 percent or greater for reasons of maintaining the continuity of service of the bulk electric power supply system.

Voltage Regulation

The maintenance of a voltage level between two established set points, compensating for transformer and/or line voltage deviation, caused by load current. The voltage change is affected by the magnitude and the power factor of the load current.

Voltage Regulator Down (VRD)

Voltage Regulator Down, an Intel standard for voltage regulators which are "down" on the mother board.

Voltage Regulator Module (VRM)

An Intel Standard for switching regulator modules.

Voltage Sag

Voltage Sags are momentary (typically a few milliseconds to a few seconds duration) under-voltage conditions and can be caused by a large load starting up (such as a air conditioning compressor or large motor load) or operation of utility protection equipment. Sags often appear as flickering lights and can cause equipment shutdown. A sag of just a few milliseconds can mean a complete blackout to some sensitive equipment.

Voltage Source

A source which essentially maintains the source voltage at a predefined value almost independent of the load conditions. In other words the terminal voltage is maintained close to the internal emf.

Voltage Spread

The difference between maximum and minimum voltages.

Voltage Swells

Voltage Swells are momentary (typically a few milliseconds to a few seconds duration) over-voltage conditions which can be caused by such things as a sudden decrease in electrical load or a short circuit occurring on electrical conductors. Voltage swells can affect the performance of sensitive electronic equipment, cause data errors, produce equip-

ment shutdowns, and may cause equipment damage.

Voltage Transducer

A transducer used for the measurement of a.c. voltage.

Voltage Transformer

Transformer used to accurately scale ac voltages up or down, or to provide isolation. Generally used to scale large primary or bus voltages to usable values for measuring purposes

Voltage Transformer Ratio

The ratio of primary volts divided by secondary volts

Voltage Transients

A transient (sometimes called impulse) is an extremely fast disturbance (millionths of a second to a few milliseconds) evidenced by a sharp change in voltage. Transients can occur on your electric, phone, or even cable TV lines. They can be caused by such things as lightning, trees falling on power lines, ice and snow, and cycling equipment ON and OFF. Transients can originate from inside or outside your home. Equipment, such as air conditioning, pump motors, photocopiers, and even electric hand tools can all cause transients when cycled on and off. These impulses are similar to lightning strikes but are much smaller. However, because they can occur frequently, they can slowly cause electronic equipment to break down.

Voltage Withstand Test

A field or factory test in which a conductor or electrical equipment is subjected to a higher than normal AC or DC voltage to test its insulation system.

Voltage, cutoff

Voltage at the end of useful discharge.

Voltage, end-point

Cell voltage below which the connected equipment will not operate or below which operation is not recommended.

Voltage-Controlled Oscillator

An oscillator device in which output frequency is proportional to its input voltage.

Voltampere (VA)

The basic unit of Apparent power. The voltamperes of an electric circuit is the mathematical product of the volt and ampere of the circuit. The practical unit of Apparent power is kilovoltampere (kVA).

Voltmeter

A device for measuring the voltage difference between any two points in an electrical circuit.

Volts

The "pressure" of electrical flow. Can be compared to the pounds per square inch (psi) of water flowing through a pipe.

Volumetric Wires Charge

A type of charge for using the transmission and/or distribution system that is based on the volume of electricity that is transmitted.

V-O-M meter

Volt-ohm-milliammeter, the troubleshooters" basic testing instrument.

Wafer

Semiconductor manufacturing begins with a thin disk of semiconductor material, called a "wafer." A series of processes defines transistors and other structures, interconnected by conductors to build the desired circuit.

The wafer is then sliced into "dice" which are mounted in packages, creating the IC.

Wafer Fab

Semiconductor processing facility which turns wafers into integrated circuits. A typical wafer fab employs a series of complex steps to define conductors, transistors, resistors, and other electronic components on the the semiconductor wafer. Imaging steps define what areas will be affected by subsequent physical and chemical processes.

Wagner Earthing

A null-balance method of keeping the arms of a bridge at earth potential without directly earthing any part of the bridge.

Wallplate

A plate designed to enclose an electrical box, with or without a device installed within the box.

Waste-to-Energy

This is a technology that uses refuse to generate electricity. In mass burn plants, untreated waste is burned to produce steam, which is used to drive a steam turbine generator. In refuse-derived fuel plants, refuse is pre-treated, partially to enhance its energy content prior to burning.

Watchdog

A feature of a microprocessor supervisory circuit that

monitors software execution in a microprocessor or microcontroller. It takes appropriate action (assert a reset or nonmaskable interrupt) if the processor gets stuck in an infinite execution loop.

Water Heating

Energy Efficiency program promotion to increase efficiency in water heating, including low-flow showerheads and water heater insulation wraps. Could be applicable to residential, commercial, or industrial consumer sectors.

Waterproof

Constructed or protected so that exposure to the weather will not interfere with successful operation.

Watertight

So constructed that water/ moisture will not enter the enclosure under specified test conditions.

Watt

1) With ac measurements, effective power (measured in Watts) equals the product of voltage, current, and power factor (the cosine of the phase angle between the current and the voltage). Watts=EI cosine(Theta). A Watt is a unit of power that considers both volts and amps and is equal to the power in a circuit in which a current of one ampere flows across a potential difference of one volt. 2) One joule/second.

Watt-Hour

One watt of power expended for one hour.

Wattmeter

An instrument for measuring the average power.

Waveform

The graphic representation of the variation of a quantity (such as voltage) as a function of some variable, usually time.

Weatherproof

A connector specially constructed so that exposure to weather will not interfere with its operation.

Weber (Wb)

SI unit of magnetic flux. One weber is equal to the magnetic flux which, linking a circuit of one turn, would produce in it an electromotive force of 1 V if it were reduced to zero at a uniform rate of 1 s.

Weir

A low dam over which a river flows, but which allows nearly all water to be diverted to a water turbine if the water level drops below its height. Helps insure year-round power production in areas with highly variable water levels.

Wells to Pump (WTP)

The first part of WTW—the part that is often overlooked when comparing different automotive technologies.

Wells to Wheels

Many comparisons between traditional gasoline-powered vehicles and alternatives look only at effects of the vehicle itself, and ignore the effects of producing fuel for the vehicle. This is remedied by Wells to Wheels calculations that include all relevant steps required to provide motive force to a vehicle, including:
extraction of raw materials (e.g. petroleum or coal) transportation (getting the raw materials to processing facilities (e.g. oil tankers or railroad)
processing (e.g. refining or electric power production) more transportation (e.g. delivery of gasoline to the pump or electric power transmission) vehicle efficiency from its fuel, which may include fuel storage efficiency conversion of the fuel into power in the engine efficiency of getting engine power to the wheels (e.g. vehicle transmission systems)
Wells to Wheels calculations fairly address the bogus complaint that EVs only move pollution problems rather than solve them. Yes, electric power plants pollute, but then so do refineries. Only by comparing the entire fuel chain can an intelligent comparison be made. By almost every measure, EVs come out ahead. Argonne National Laboratories, a publicly-funded US Research Lab, has developed a public domain spreadsheet model to profile greenhouse gas emissions and net energy usage for various transportation modes. The Vancouver Electric Auto Association has an excellent summary of energy efficiency studies completed by Argonne National Laboratories.

Western Systems Coordinating Council (WSCC)

One of the ten regional reliability councils that make up the North American Electric Reliability Council (NERC).

Western Underground Committee

A committee of western based electric utility engineers that provides a forum for establishment of guides that provide options, recommendations and practices for its members. These guides are used to assist its members in preparing their own specifications and to make recommendations to specificying agencies.

Wet Cell

A cell, the electrolyte of which is in liquid form and free to flow and move.

Wet Shelf Life

The period of time that a charged battery, when filled with electrolyte, can remain unused before dropping below a specified level of performance.

Wheastone Bridge

A four arm resistance bridge where the balance is obtained by null deflection when the ratio of the adjacent arms are equal.

Wheeling

Transmitting bulk electricity from a generating plant to a distribution system across a third system's lines.

Wheeling Service

The movement of electricity from one system to another over transmission facilities of intervening systems. Wheeling service contracts can be established between two or more systems.

Whole House Surge Suppressor

A type of surge suppressor that is designed to protect the whole house from power surges originating on the utility distribution system. Separate protection devices are required for telephone and cable systems to protect against surges traveling through the telephone or cable.

Whole-House Fan

A large fan used to ventilate your entire house. This is usually located in the highest ceiling in the house, and vents to the attic or the outside. Although whole-house fans are a good way to draw hot air from the house, you must be careful to cover and insulate them during the winter, when they often continue to draw hot air from people's houses.

Wholesale Bulk Power

Very large electric sales for resale from generation sources to wholesale market participants and electricity marketers and brokers.

Wholesale Competition

A system whereby a distributor of power would have the option to buy its power from a variety of power producers, and the power producers would be able to compete to sell their power to a variety of distribution companies.

Wholesale Power Market

The purchase and sale of electricity from generators to resellers (who sell to retail consumers) along with the ancillary services needed to maintain reliability and power quality at the transmission level.

Wholesale Sales

Energy supplied to other electric utilities, cooperatives, municipals, and Federal and State electric agencies for resale to ultimate consumers.

Wholesale Transition

The sale of electric power from an entity that generates electricity to a utility or other electric distribution system through a utility's transmission lines.

Wholesale Transmission Services

The transmission of electric energy sold, or to be sold, at wholesale in interstate commerce (from EPACT).

Wide Area Network (WAN)

Any Internet or network that covers an area larger than a single building.

Wideband

A classification of the information capacity or bandwidth of a communication channel. Wideband is generally understood to mean a bandwidth between 64kbits/s and 2Mbit/s.

Wideband Code Division Multiple Access (WB-CDMA)

A standard derived from the original CDMA. WB-CDMA is the third-generation (3G) mobile wireless technology capable of supporting voice, video, and data communications up to 2Mbps.

Wind Energy Conversion
A process that uses energy from the wind and converts it into mechanical energy and then electricity.

Wind Generator
A device that captures the force of the wind to provide rotational motion to produce power with an alternator or generator.

Wind Power Terminology
Terms that apply specifically to wind turbines and wind energy systems.

Wind Turbine
A machine that captures the force of the wind. Called a Wind Generator when used to produce electricity. Called a Windmill when used to crush grain or pump water.

Windmill
Devices that uses wind power to mill grain into flour. Informally used as a synonym for wind generator or wind turbine, and to describe machines that pump water with wind power.

Window
A wide band gap material chosen for its transparency to light. Generally used as the top layer of a photovoltaic device, the window allows almost all of the light to reach the semiconductor layers beneath.

Window Comparator
A device, usually consisting of a pair of voltage comparators, in which output indicates whether the measured signal is within the voltage range bounded by two different thresholds (an "upper" threshold and a "lower" threshold).

Window Watchdog
A special subset of the watchdog timer feature found on microprocessor supervisory circuits. It is used to monitor software execution and assert a reset or an NMI if the processor gets stuck in a loop. This feature not only looks for periodic transitions on its input within a preprogrammed timeout period, but it also looks to see if there are "too many" transitions within its timeout period (window).

Windward

Toward the direction from which the wind blows.

Winston Concentrator

A trough-type parabolic collector with one-axis tracking, developed by Roland Winston.

Winter Peak

The greatest load on an electric system during any prescribed demand interval in the winter season or months.

Wire

A strand or group of strands of electrically conductive material, normally copper or aluminum.

Wire Lubricant

A chemical compound used to reduce pulling tension by lubricating a cable when pulled into a duct or conduit.

Wireless

Radio-frequency devices, circuits, or communications methods.

Wireless Bridging

A networking bridge is used to connect two or more separate networks. A wireless bridge functions similar to a wireless network but can be used in situations in which running a cable would be impractical or expensive.

Wireless Local Area Network (WLAN)

A wirelessly connected Local Area Network.

Wireless Local Loop (WLL)

Any method of using wireless communication in place of a wired connection to provide subscribers with standard telephone service.

Wireless Personal-Area Network (WPAN)

WPAN is a PAN that uses wireless means of connecting, however since all PAN technologies, such as Bluetooth, are wireless, you can consider the terms synonymous.

Wireless Sensor Network

Wireless Sensor Network, or WSN, is a network of RF transceivers, sensors, machine controllers, microcontrollers, and user interface devices with at least two nodes communicating by means of wireless transmissions.

Wireless Telephony Application (WTA)

A collection of telephony-specific extensions for call- and feature-control mechanisms that make advanced mobile network services available to end users. WTA essentially merges the features and services of data networks with the services of voice networks.

Wires Charge

A broad term which refers to charges levied on power suppliers or their customers for the use of the transmission or distribution wires.

Wiring

A distribution network of wire that conducts electricity to receptacles, switches and appliances throughout a building/home to provide electricity where needed.

Wiring System

An assembly made up of cable or busbars and parts which secure and, if necessary, enclose the cable or busbars.

Word

A group of bits representing a complete piece of digital information.

Work Function

The energy difference between the Fermi level and vacuum zero. The minimum amount of energy it takes to remove an electron from a substance into the vacuum.

Work Plane

The plane at which work is usually done and on which the illuminance is specified and measured. Unless otherwise indicated, this is assumed to be a horizontal plane 30" above the floor.

Working Near

Refers to working near live parts. Any activity inside a limited approach boundary.

Working On

Refers to working on live parts. Coming in contact with live parts with the hands, feet, or other body parts, with tools, probes, or test equipment, regardless of the personal protective equipment a person is using.

Working Voltage (WV)

The maximum voltage that may be applied continuously to a capacitor without risking breakdown.

Wound Rotor

Rotor of an induction motor provided with a three-phase winding in the rotor.

Wye

A three phase, four-wire electrical configuration where each of the individual phases is connected to a common point, the "center" of the Y. This common point normally is connected to an electrical ground.

Wye-Delta Starting

Wye-delta is a connection which is used to reduce the inrush current and torque of a three-phase motor. A wye (star) start, delta run motor is one arranged for starting by connecting to the line with the winding initially connected wye (star). The winding is then reconnected to run in delta after a predetermined time. The lead numbers for a single run voltage are normally 1, 2, 3, 4, 5 and 6.

X-rays

An electromagnetic radiation produced when the inner satellite electrons of heavy atoms have been excited by collision with a stream of fast electrons return to their ground state, giving up the energy previously imparted to them. Electromagnetic radiations of the same type as light, but of much shorter wavelength, in the range of 5 nm to 6 nm produced when cathode rays strike a material object.

X-Y Plot

A graphic representation of the relationship of the X signal, which controls the horizontal position of the beam in time, and the Y signal, which controls the vertical position of the beam in time.

Yaw

Rotation parallel to the ground. A wind generator Yaws to face winds coming from different directions.

Yaw Axis

Vertical axis through the center of gravity.

Yield Strength

The stress required to produce a very slight yet specified amount of plastic strain.

yotta (Y)

Decimal multiple prefix corresponding to 1024

Young's modulus

Elastic modulus applied to a stretched wire or to a rod under tension or compression. The ratio of the stress to the strain.

Yttrium-Iron-Garnet

Yttrium-iron-garnet (YIG) is a ferrimagnetic material used for solid-state lasers and for microwave and optical communications devices.

Z-axis

Refers to the signal in an oscilloscope that controls electron-beam brightness as the trace is formed.

Zener Diode

A junction diode designed to operate in the reverse bias region.

Zenith Angle

The angle between the direction of interest (of the sun, for example) and the zenith (directly overhead).

Zepto (Z)

Decimal sub-multiple prefix corresponding to 10-21

Zero Crossing

The point at which a sinsoidal voltage or current waveform crosses the zero reference axis.

Zero Emissions Vehicle (ZEV)

ZEVs have zero tailpipe emissions are 98% cleaner than the average new model year vehicle. These include battery electric vehicles and hydrogen fuel cell vehicles.

Zero Insertion Force (ZIF)

A class of IC sockets which clamp the IC pins (via a small lever on the side of the socket) after insertion, and thus require no downward force on the IC or its pins to insert it into the socket. Especially useful in applications in which repeated insertions subjects the IC or the socket to wear and breakage.

Zero Sequence

A balanced set of three phase components which have the same magnitude and the same phase angle, and hence hence no sequence. The frequency is of course the same as the original unbalanced three phase system.

Zero Signal Reference

A connection point, bus, or conductor used as one side of a signal circuit. It may or may not be designated as ground. Is sometimes referred to as circuit common.

Zetta (Z)

Decimal multiple prefix corresponding to 1021
The signal in an oscilloscope that controls the electron-beam brightness as the trace is formed.